T0205514

Agricultural Biotechnology in Sub-Saharan Africa

John Edward Otieno Rege • Keith Sones

Editors

Agricultural Biotechnology in Sub-Saharan Africa

Capacity, Enabling Environment and Applications in Crops, Livestock, Forestry and Aquaculture

 Springer

Editors
John Edward Otieno Rege
Emerge Centre for Innovations – Africa
Nairobi, Kenya

Keith Sones
Keith Sones Associates
Banbury, UK

ISBN 978-3-031-04351-2   ISBN 978-3-031-04349-9   (eBook)
https://doi.org/10.1007/978-3-031-04349-9

This Springer imprint is published by the registered company Springer Nature Switzerland AG
The registered company address is: Gewerbestrasse 11, 6330 Cham, Switzerland

Acknowledgements

The editors would like to thank the Food and Agriculture Organization of the United Nations (FAO) for providing the opportunity to write this book. The book builds upon the material originally generated for a background study on agricultural biotechnology for the Sub-Saharan Africa region, commissioned by FAO, which was undertaken by the Emerge Centre for Innovations—Africa (formerly, the Institute for People, Innovation and Change in Organizations—Eastern Africa or PICO Eastern Africa), led by Ed Rege. In turn, this study built on an international symposium, The Role of Agricultural Biotechnologies in Sustainable Food Systems and Nutrition, convened by the FAO and held in February 2016 at FAO headquarters, Rome.

About this Book

The starting point for this book was an international symposium convened by the Food and Agriculture Organization of the United Nations (FAO), The Role of Agricultural Biotechnologies in Sustainable Food Systems and Nutrition, held in February 2016 at FAO headquarters, Rome. The symposium attracted delegates from 75 FAO member countries and the European Union, as well as representatives of intergovernmental organisations, the private sector, civil society, research and academia, and farmers'/producers' organisations and cooperatives.

The Symposium highlighted the important contribution that agricultural biotechnologies make or can potentially make in achieving the Sustainable Development Goals. The importance of bringing the dialogue from the global to the regional level was highlighted by participants. In response, the FAO is committed to bringing the conversation to the regions in order "to hear from governments, farmers and researchers of all regions about their needs and concerns regarding biotechnology".

FAO undertook to commission and publish a regional background study to form a basis for each regional conversation. The Emerge Centre for Innovations—Africa (formerly, the Institute for People, Innovation and Change in Organizations—Eastern Africa or PICO Eastern Africa) was commissioned to undertake the study for the Sub-Saharan Africa (SSA) region. Having done so, the authors considered that it would be useful to extend the scope of that study and further develop, organise, and update the material. The result is this book.

Four types of source material were used to write this book. The primary source was a systematic literature review. This was supplemented by questionnaire-based surveys and key informant interviews. Two surveys were undertaken. The first was undertaken by the FAO, who sent questionnaires to senior government officials responsible for crops, livestock, forestry, and fisheries/aquaculture in all (SSA) countries in May/June 2017. The second was an online survey undertaken by the authors in June/July 2017. For this, 1866 individuals from 44 SSA countries—researchers, research managers, or people otherwise involved in other relevant aspects of agricultural development—were invited to complete the online survey. In addition, key informant interviews were used when specific pieces of information were needed to fill gaps or address queries. In this case, individuals were identified and approached via telephone, messaging apps, or e-mail. These informants suggested additional reference material.

Contents

Editors and Contributors

About the Editors

J. E. O. Rege is founder and chief executive officer of Emerge Centre for Innovations—Africa (ECI-Africa), a not-for-profit organisation that supports institutional development in the areas of agriculture and rural development. He has over 35 years' experience in agriculture, having worked in different capacities—as university trainer, CGIAR researcher, and science leader. Ed holds a PhD in genetics, MS in animal science, and BSc in agriculture. Prior to founding ECI-Africa, Ed was the director of the Global Biotechnology Program of the International Livestock Research Institute.

Keith Sones is a freelance writer and editor, specialising in African agriculture and livestock disease, and works for both public and private sector organisations. Currently based in rural Oxfordshire, UK, he lived in Nairobi, Kenya, for 30 years, initially working for a major veterinary pharmaceutical company before becoming a freelance writer and editor, and workshop and conference organiser, facilitator, and rapporteur mainly in the livestock sector. Keith holds a PhD in parasitology from Glasgow Veterinary School, Scotland.

Contributors

Dionysious Kiambi African Biodiversity Conservation and Innovations Centre, Nairobi, Kenya

Charles Midega Poverty and Health Integrated Solutions (PHIS), Kisumu, Kenya

Joel W. Ochieng Agricultural Biotechnology Programme, University of Nairobi, Nairobi, Kenya

Abbreviations

AATF	African Agricultural Technology Foundation
ABCF	Africa Biosciences Challenge Fund
ABI	African Biosciences Initiative
ABSF	African Biotechnology Stakeholders Forum
ACE	Africa Higher Education Centers of Excellence
ADGG	African Dairy Genetics Gains
AFF	African Forest Forum
AFORNET	African Forest Research Network
AGRA	Alliance for a Green Revolution in Africa
AI	artificial insemination
AnGR	animal genetic resources
APPSA	Agricultural Productivity Program for Southern Africa
ARC	Agricultural Research Centre, South Africa
ARC	Agricultural Research Corporation, Sudan
ARD	agricultural R&D
ASARECA	Association for Strengthening Agricultural Research in Eastern and Central Africa
ASTI	Agricultural Science and Technology Indicators
AU	African Union
AUC	African Union Commission
AU-PANVAC	Pan African Veterinary Vaccine Centre
AU-IBAR	The African Union–Interafrican Bureau for Animal Resources
BecA	Biosciences Eastern and Central Africa
BIC	Biotechnology Information Center
BioAWARE	National Biotechnology Awareness Creation Strategy, Kenya
BioEROC	Biotechnology-Ecology Research and Outreach Consortium
BMGF	Bill & Melinda Gates Foundation
BOSZ	Biotechnology Outreach Society of Zambia
BSL	biosafety level
Bt	*Bacillus thuringiensis*
BV	breeding value
BVI	Botswana Vaccine Institute
CAADP	Comprehensive African Agricultural Development Programme

CAR	Central African Republic
CBBPs	community-based breeding programmes
CBD	Convention on Biological Diversity
CCARDESA	Centre for Coordination of Agricultural Research and Development for Southern Africa
CDT	conventional drought tolerant
CIDA	Canadian International Development Agency
CIP	International Potato Center
CGIAR	formerly the Consultative Group for International Agricultural Research
CIMMYT	International Maize and Wheat Improvement Center
CIRAD-EMVT	Centre de Cooperation Internationale en Recherche Agronomique pour le Developpement- Departement d'Elevage et de Medecine Veterinaire
CIRDES	Centre international de recherche-développement sur l'elevage en zone subhumide
CIFOR	Centre for International Forestry Research
CNRA	Centre National de Recherche Agronomique, Côte d' Ivoire
COMESA	Common Market for Eastern and Southern Africa
CORAF	Conseil Ouest et Centre Africain pour la Recherche et le Développement Agricoles
CRISPR	Clustered regularly interspaced short palindromic repeats
CSIR	Council for Scientific and Industrial Research, Ghana
CTTBD	Centre for Ticks and Tick-borne Diseases
DarTs	Diversity arrays technology
DNA	Deoxyribonucleic acid
DRC	Democratic Republic of the Congo
EAAPP	East African Agricultural Productivity Program
EAC	East African Community
ECA	Economic Commission for Africa
ECAAT	Eastern and Central Africa Agriculture Transformation
ECF	East coast fever
ECOWAS	Economic Community of West African States
ELISA	Enzyme-linked immunosorbent assay
EMBRAPA	Empresa Brasileira de Pesquisa Agropecuária (Brazilian Agricultural Research Corporation)
ET	Embryo transfer
FABI	Forestry and Agricultural Biotechnology Institute
FAO	Food and Agriculture Organization of the United Nations
FARA	Forum for Agricultural Research in Africa
FGR	Forest genetic resources
FMD	Foot-and-mouth disease
FOFIFA	National Center for Applied Research on Rural Development, Madagascar

FORIG	Forestry Research Institute of Ghana
FORNESSA	Forestry Research Network for Sub-Saharan Africa
FRIN	Forestry Research Institute of Nigeria
FTE	Full-time equivalent staffing
GALVmed	Global Alliance for Livestock Veterinary Medicines
GDP	Gross domestic product
GM	Genetically modified
GMO	Genetically modified organism
HAT	Human African trypanosomiasis
IAEA	International Atomic Energy Agency
IBAR	Interafrican Bureau for Animal Resources
ICAR	International Committee for Animal Recording
ICARDA	International Center for Agriculture Research in the Dry Areas
ICFR	Institute for Commercial Forestry Research, South Africa
ICIPE	International Centre of Insect Physiology and Ecology
ICRAF	World Agroforestry Centre
ICRISAT	International Crops Research Institute for the Semi-Arid Tropics
IETS	International Embryo Transfer Society
IFAD	International Fund for Agricultural Development
IFPRI	International Food Policy Research Institute
IIAM	Agricultural Research Institute of Mozambique
IITA	International Institute of Tropical Agriculture
ILRI	International Livestock Research Institute
IPMS	Improving the Productivity and Market success of Ethiopian Farmers
ISAAA	International Service for the Acquisition of Agri-biotech Applications
ITM	Infection and treatment method
IVF	In vitro fertilisation
KALRO	Kenya Agriculture and Livestock Research Organization
KARI	Kenya Agriculture Research Institute
KEFRI	Kenya Forestry Research Institute
KEVEVAPI	Kenya Veterinary Vaccines Production Institute
LCV	Central Veterinary Laboratory, Mali
LIVES	Livestock and irrigation value chains for Ethiopian smallholders
LMO	Living modified organism
LUANAR	Lilongwe University of Agriculture and Natural Resources
MAS	Marker-assisted selection
MOET	Multiple ovulation embryo transfer
NABNet	North Africa Biosciences Network
NAIC	National Artificial Insemination Center, Ethiopia
NAPRI	National Animal Production Research Institute
NARI	National Agricultural Research Institute

NARO	National Agricultural Research Organisation
NARS	National Agricultural Research System
NBF	National Biosafety Framework
NEPAD	New Partnership for Africa's Development
NGO	Non-governmental organisation
OFAB	Open Forum on Agricultural Biotechnology
OFSP	Orange-fleshed sweet potato
OIE	World Organisation for Animal Health
OS	Oestrus synchronisation
OVI	Onderstepoort Veterinary Institute
PABRA	Pan-Africa Bean Research Alliance
PAID	Public Private partnership for AI Delivery
PCR	Polymerase chain reaction
PPP	Public–private partnership
PUB	Public Understanding of Biotechnology, South Africa
QTL	Quantitative trait loci
R&D	Research and development
RAPD	Random amplified polymorphic DNA marker
RIA	Radioimmunoassay
RARI	Regional Agricultural Research Institutes, Ethiopia
REC	Regional economic community
RECOAB	Réseau des Communicateurs ouest-Africains sur la Biotechnologie
RNAi	RNA interference
SADC	Southern African Development Community
SAFORGEN	Sub-Saharan African Forest Genetic Resources
SANBio	Southern Africa Network for Biosciences
SCNT	Somatic cell nuclear transfer
SIFET	Sexed in vitro fertilisation embryo transfer
SIT	Sterile insect technique
SNP	Single-nucleotide polymorphism
SoW-FGR	FAO's State of the World's Forest Genetic Resources
SPGS	Sawlog Production Grant Scheme
SSA	Sub-Saharan African
SSR	Simple sequence repeat
TBPT	Tree Biotechnology Project Trust
TC	Tissue culture
UNEP-GEF	United Nations Environment Programme-Global Environment Facility
USAID	United States Agency for International Development
USD	United States dollar
VACNADA	Vaccines for the Control of Neglected Animal Diseases in Africa
WAAPP	West Africa Agricultural Productivity Program

WABNet	West Africa Biosciences Network
WECARD	West and Central African Council for Agricultural Research and Development
WEMA	Water Efficient Maize for Africa

The Agriculture Sector in Sub-Saharan Africa and the Promise of Biotechnology

1

J. E. O. Rege and Keith Sones

Abstract

Agriculture supports the livelihoods of most households in sub-Saharan Africa (SSA) and makes significant contribution to national economies, especially in countries that are not dependent on mineral wealth. The region is home to more than 950 million people and is projected to reach 2.1 billion by 2050. Although agriculture accounts for about 23% of the region's total gross domestic product, the full potential of the agriculture sector in SSA is yet to be exploited. Food insecurity on the continent has been increasing since 2014: in 2019, 250 million people in SSA were undernourished, and the coronavirus pandemic has likely pushed tens of millions more into food insecurity. Climate change will exacerbate existing threats to food security and agriculture-based livelihoods. Africa's 33 million smallholder farmers depend on rainfed agriculture and are especially vulnerable to impacts of climate change. Unlike other regions of the world, where yields have increased substantially over recent decades, yields have stagnated in SSA. The African Union's target of 6% annual agricultural productivity growth will require substantial policy support and investment in agricultural technologies, including biotechnology. With the exception of South Africa, most SSA countries continue to question whether agricultural biotechnology is a good investment.

J. E. O. Rege (✉)
Emerge Centre for Innovations-Africa, Nairobi, Kenya
e-mail: ed.rege@emerge-africa.org

K. Sones
Keith Sones Associates, Banbury, UK

Table 1.1 SSA countries where agriculture contributes

More than 20% to GDP (%)	Less than 10% to GDP (%)
Benin (26.9)	Angola (6.7)
Burkina Faso (20.2)	Botswana (1.9)
Burundi (28.9)	Cabo Verde (4.8)
Central African Republic (32.4)	Congo Republic (7.8)
Chad (42.6)	Equatorial Guinea (2.5)
Comoros (33.1)	Eswatini[a] (8.5)
Côte d' Ivoire (20.7)	Gabon (5.6)
Ethiopia (33.5)	Lesotho (4.5)
The Gambia (21.8)	Mauritius (2.9)
Guinea (23.6)	Namibia (6.6)
Guinea-Bissau (52.5)	Seychelles (2.3)
Kenya (34.1)	South Africa (1.9)
Liberia (39.1)	Zambia (2.9)
Madagascar (23.3)	Zimbabwe (8.3)
Malawi (25.5)	
Mali (37.3)	**10–20% to GDP (%)**
Mozambique (26.0)	Cameroon (14.5)
Niger (37.8)	D R Congo (20.0)
Nigeria (21.9)	Eritrea (14.1)
Rwanda (23.5)	Ghana (17.3)
Sierra Leone (54.3)	Mauritania (18.7)
Somalia (62.7)	Senegal (14.8)
Tanzania (28.7)	Sao Tome and Principe (12.5)
Togo (22.5)	South Sudan (10.5)
Uganda (23.1)	Sudan (11.6)

For countries in **bold**, agriculture contributes more than 50% to GDP
[a]Formerly Swaziland
Source: World Bank National Accounts Data and OECD National Accounts Data files (World Bank 2021)

1.1   Agriculture Sector in Sub-Saharan Africa

In sub-Saharan Africa (SSA), agriculture supports the livelihoods of 90% of the people, represents just under 16% of the region's gross domestic product (GDP) and accounts for almost 8% of the continent's export income (worth around USD 40 billion a year) (African Union 2020). However, the contribution of agriculture to total GDP ranges from below 3% in Botswana and South Africa to 54% in Sierra Leone and 62% in Somalia, implying a diverse range of economic structures. As of 2019, the contribution of agriculture to GDP was 20% or more in 25 (52%) of SSA countries, while the sector contributed less than 10% to GDP in 13 (27%) of the 48 countries (Table 1.1), most of which are dependent on mineral wealth (World Bank 2021). The declining GDP contribution of agriculture relative to other sectors of the economies of African countries is attributable more to low productivity and limited value addition to agricultural commodities than it is an indication of low potential of agriculture.

The period 1980 to 2005 delivered unparalleled progress and improvements in quality of life across the developing world. Agricultural growth in SSA rose from an annual rate of 2.3% in the 1980s to an average of 3.8% for the years 2000 to 2005. This was good news as it is estimated that rural poverty is reduced by more than 1.8% for every 1% increase in agricultural growth (Fan et al. 2008). No other sector has more impact on rural poverty. However, the agricultural growth that Africa experienced was mostly based on expanding land use. Productivity hardly increased. This has to change because agriculture has remained critical to the survival of the majority of the population in SSA, and food insecurity has worsened in recent years. According to the Food and Agriculture Organization of the United Nations (FAO), the prevalence of undernourishment in Africa was 19.1% (more than 250 million people) in 2019, up from 17.5% in 2014. The impacts of the COVID-19 pandemic are likely to move tens of millions more into food insecurity. Even without allowing for COVID-19, it is projected that by 2030 Africa will become the region with the highest number of undernourished people (433 million), accounting for more than half of the global total (FAO et al. 2020). The continent's population was 1 billion in 2009 and is expected to rise to 2.2 billion by 2050, thus worsening the situation.

Climate change will act as a multiplier of existing threats to food security: by 2050, the risk of hunger and child malnutrition is anticipated to be 10–20% higher compared to a 'no-climate change' scenario. Africa's 33 million smallholder farmers, who make up 70% of the population (AGRA 2017) and produce up to 70% of the food supply, are the most adversely affected by climate change. This is because agriculture in SSA is predominantly rainfed; very few farms benefit from irrigation. The situation is worsened by poor access to technologies, low soil fertility, weak land-tenure systems, a lack of access to finance and a lack of transport and market infrastructure, among other challenges. Together, these present a pre-existing 'development deficit' that demands special attention even in a 'no-climate change scenario' and significantly increases the vulnerability of African agriculture to climate change.

1.2 Agricultural Research and Development in Sub-Saharan Africa

The continent's agricultural development is in a race against time to eliminate this deficit while simultaneously adapting to a rapidly changing climate. The stakes are extraordinarily high as climate change is expected to lead to significant reductions in crop yields, threatening the livelihoods of hundreds of millions of poor subsistence farmers and agricultural workers (Bailey et al. 2014). Under the worst-case climate change scenario, by the middle of this century, a reduction in mean yield of 13% is projected in West and Central Africa and 8% in eastern and southern Africa (WMO 2020).

Closing the development deficit, and providing farmers with access to the invest-ment, technologies and knowledge they need to adapt to climate change, could transform their development prospects. Sustainably increasing farm productivity is therefore a priority as yields have stagnated at levels well below global averages. Narrowing the yield gap could increase farm incomes and food availability in Africa, with beneficial impacts on hunger and poverty.

Agricultural research, including those addressing climate adaptation—e.g. drought-resistant crops and livestock and ecological restoration interventions—can potentially play a critical role in breaking the vicious cycle of poverty and food insecurity in Africa, enabling poor farmers to increase their productivity and on-farm food production. Higher productivity, especially of food staples, in turn creates a ripple effect throughout the economy by lowering food prices for poor consumers, which raises real incomes. This, in turn, would benefit not only the urban poor but also food deficit farmers and rural households who constitute most of the poor in Africa.

Higher real incomes among the poor would allow them to consume more nutritious foods and enable them to access a range of other goods and services such as housing, education and healthcare, thus improving welfare of these households. This would provide stimulus for non-farm growth (Lipton 2001). In addition, higher incomes empower the poor as it enhances opportunities for collec-tive action (Garrity 2002). Moreover, because food staples are the main source of nutrients in the diets of the rural poor, productivity-improving research improves health of the poor through greater consumption, not to mention benefits accruing from research that directly addresses nutrient density of staples.

Globally, biotechnology is offering unprecedented opportunities for increasing agricultural productivity, contributing to natural resource management and protecting the environment through reduced use of agrochemicals. However, the major thrust in biotechnology research is currently directed at solving immediate problems of industrialized countries with major investments coming from multina-tional companies. Nevertheless, it is increasingly clear that many of the new discoveries and products will find their biggest markets in developing countries where the potential for improvement in agricultural productivity and human health is greatest.

Sub-Saharan Africa's agriculture has not benefitted significantly from the tech-nological innovations of the last half-century. For instance, in other regions of the world, per hectare crop yields have doubled or even quintupled, but yields have not increased substantially in Africa (Bjornlund et al. 2020). Scientific and technological advances could be used to mitigate some of the factors that continue to keep African agricultural productivity at extremely low levels. Prospects exist for significant productivity improvement through a combination of technological and policy measures.

The African Union's (AU) Agenda 2063 aspiration is for a 'prosperous Africa based on inclusive growth and sustainable development'. The AU's Comprehensive African Agricultural Development Programme (CAADP), one of the continental frameworks under Agenda 2063, aims to help African countries eliminate hunger and reduce poverty by raising agriculture-led economic growth. To achieve this, African governments have committed to increase their investment level in agriculture by allocating at least 10% of national budgets to agriculture and rural development to achieve agricultural growth rates of at least 6% per annum (AUC 2014).

Realizing a 6% agricultural productivity growth rate is possible, but it will need unprecedented commitment to policy development and practice by African governments and international development partners. Such policy shifts should aim for sustained investment in the next generation of agricultural technologies, including biotechnologies. They also need to focus on Africa's objectives for poverty reduction and on the drive to ensure that both African agriculture and its economic growth are sustainable (Brink et al. 1998).

1.3   Agricultural Biotechnology

The Convention on Biological Diversity (CBD) defines biotechnology as 'any technological application that uses biological systems, living organisms, or derivatives thereof, to make or modify products for specific use' (CBD 1992). This definition includes medical and industrial applications as well as many of the tools and techniques that are commonplace in agriculture and food production. The FAO Glossary of Biotechnology defines biotechnology broadly as in the CBD and narrowly as 'a range of different molecular technologies such as gene manipulation and gene transfer, DNA typing and cloning of plants and animals' (FAO 2001). The US Department of Agriculture has defined agricultural biotechnology as 'a range of tools, including traditional breeding techniques, that alter living organisms, or parts of organisms, to make or modify products, improve plants or animals, or develop microorganisms for specific agricultural uses' (USDA 2016).

Recombinant DNA techniques, also known as genetic engineering, refer to the modification of an organism's genetic make-up using transgenesis, in which DNA from one organism or cell (the transgene) is transferred to another without sexual reproduction. Genetically modified organisms (GMOs) are modified by the application of transgenesis or recombinant DNA technology, in which a transgene is incorporated into the host genome or a gene in the host is modified to change its level of expression. Thus, biotechnology is not the same as genetic engineering. Indeed, some of the least controversial aspects of agricultural biotechnology are potentially the most powerful and the most beneficial for the poor. Genomics, for example, is revolutionizing our understanding of the ways genes, cells, organisms and ecosystems function and is opening new horizons for marker-assisted breeding and genetic resource management. At the same time, genetic engineering is a powerful tool and one whose role should be carefully evaluated. Importantly, it is critical to understand how biotechnology, especially genetic engineering, can complement other approaches and tools if sensible decisions are to be made about its use.

The term 'biotechnology' is used in this book in its broadest sense, as per Article 2 of the CBD (1992), to include traditional ('low-tech'), at the one end, and modern ('high-tech') biotechnology, at the other end (see Fig. 1.1).

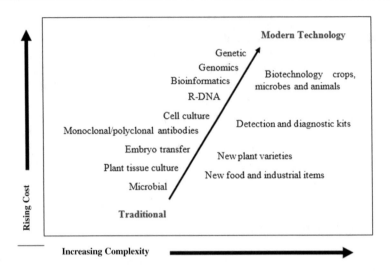

Fig. 1.1 Range of agricultural biotechnologies

No technology can be claimed to be completely risk-free. While genetically engineered crops can reduce some environmental risks associated with conventional agriculture, such as use of pesticides, new challenges will be introduced that must be addressed. Society has to decide when and where genetic engineering is safe enough. Thus, while modern biotechnologies have a role to play in improving agricultural productivity and production and in addressing some of the major environmental challenges, the aim is not and should not be to replace conventional farming practices that work and are safe. For SSA to achieve the target 6% productivity growth, it will require the deployment of appropriate mix of low-, medium- and high-tech agricultural biotechnologies, including genetic engineering. But SSA Africa cannot afford the luxury of not embracing emerging biotechnologies. The future survival of SSA countries in the highly competitive international market will depend on how well they integrate science, technology and innovation, particularly the rapidly evolving field of modern biotechnology, in national development agendas. Biotechnologies can, for example, enable faster and more efficient breeding programmes that produce new genotypes (varieties and breeds) of crops, livestock and fish that can overcome longstanding and emerging challenges from pests and diseases and the impacts of climate change. Such programmes can provide crops, meat, milk, eggs and fish and also various forest products with increased productivity, improved yield stability and quality traits, such as enhanced nutritional value, while using fewer potentially harmful and costly pesticides. Biotechnologies cannot, however, overcome failings with regard to infrastructure, markets and input and extension systems that constrain the uptake of other improved technologies and approaches.

The promise of biotechnology has led some governments in both industrialized and developing countries to view it as the next major engine of economic growth.

Major government investments have been made in medical and agricultural biotechnology research in the USA and countries in Europe as well as in emerging economies such as China, India and Brazil.

In SSA, South Africa stands out as the country that has most fully embraced biotechnology. African governments, by and large, continue to question whether agricultural biotechnology is a good investment, given the high costs of research and regulation and concerns about patents and the roles of multinational corporations. But perhaps the key blockage to widespread acceptance of agricultural biotechnology is the ongoing debate about food safety and environmental impacts, especially regarding high-end biotechnologies such as genetic modifications. Preconditions for both effective and safe research and applications of biotechnology (in any field) include having a critical mass of capacity (human and infrastructural) as well as an enabling environment (including policies, regulations and investments).

This book focuses on the capacities and enabling environments for, and status of applications of, agricultural biotechnologies in SSA countries in crop, livestock, forestry and aquaculture improvement.

1.4   The Classification Framework Applied

Chapter 2 assesses SSA countries in terms of their capacities for research and applications of agricultural biotechnology in crops and livestock. The state of capacities for biotechnology applications in forestry and aquaculture is analysed in Chaps. 6 and 7, respectively. Capacities examined include human capacities, institutions and facilities (including labs, equipment and operational budget) and networks and networking. For each sector—crops, livestock, forestry and aquaculture—countries are classified into five biotechnology capacity classes: 'very low', 'low', 'medium, 'high' and 'very high'.

Chapters 3 (for crops and livestock), Chap. 6 (forestry) and Chap. 7 (aquaculture) examine and classify countries on the basis of factors which define an enabling environment for biotechnology research and applications, namely, public awareness, participation and acceptance; national biosafety frameworks; public policy and political goodwill; and financial and investment environment. For each sector—crops, livestock, forestry and aquaculture—countries are classified into five enabling environment classes: 'very weak', 'weak', 'medium, 'strong' and 'very strong'.

Chapters 4 and 5 focus on sector assessments of the status of applications of biotechnologies in crops and livestock, respectively. Assessments of biotechnology applications in forestry and aquaculture are covered in Chaps. 6 and 7, respectively. For each sector—crops, livestock, forestry and aquaculture—countries are classified into five biotechnology application classes: 'very low', 'low', 'medium, 'high' and 'very high' use. In the case of applications, two main considerations have been used in classifying countries in these chapters: the extent of use and context of use in both research and commercial/field applications.

Extent of use Two countries may be using a particular technology but the extent—depth and breadth—of use may differ, sometimes significantly. For example, genetic engineering is applied in both Kenya and South Africa, yet South Africa is ranked 'very high use' while Kenya is ranked 'high use'. This distinction is made because, while currently genetic engineering is applied in Kenya at laboratory level (and in controlled field trials), South Africa has moved the technology to the field and farmers are applying it widely. Similarly, a technology such as the polymerase chain reaction (PCR) used in a single institution in a country for a specific purpose (e.g. in research) is ranked lower compared to a country that uses PCR routinely and widely in multiple institutions across the country for diagnostics and diversity studies and as part of genetic engineering protocols.

Context of use The context of use was considered for each technology. Gene cloning, for example, as a tool for isolating a gene from the rest of the genome so that the operator can get better quality sequences in molecular genetic studies (a medium-level application—'medium-tech'), is different from cloning to obtain an identical organism, such as the *Futhi* heifer of South Africa or Dolly (the sheep) (both 'high-tech')—see Sect. 5.2.3 in Chap. 5. In the same way, bioinformatics being applied in analysis of biochemical pathways, docking[1] and 3D drug design, or assembling next-generation sequencing data and creating genetic linkage maps is clearly being used at a level different from using molecular biology tools in phylogenetic reconstruction for conservation. The different ways in which biotechnologies are used is captured by classifying them as 'low-tech', 'medium-tech' or 'high-tech'.

The 'low-tech' applications of biotechnology as used in this book include phenotypic selection as applied in variety and breed development, development and use of bio-fertilizers (such as rhizobial inoculants for legumes), artificial insemination (AI) in livestock and sex reversal and polyploidy in farmed fish. Examples of 'medium-tech' applications are the use of the polymerase chain reaction (PCR) for deoxyribonucleic acid (DNA) marker-assisted selection in crops, livestock, forestry or aquaculture; embryo transfer (ET) in livestock; and tissue culture in crops and trees, while 'high-tech' applications include gene editing, genetic engineering and development of new varieties of crops through genetic engineering and cloning.

The definitions used to classify countries are summarized in Table 1.2. The same approach was used for crops, livestock, forestry and aquaculture. 'High-tech' applications are being used in only a few countries. Low-tech and medium-tech applications were the most common in the majority of countries. The same small number of countries using 'high-tech applications (experimentally or commercially) is generally widely, even routinely, using low- and medium-tech applications in multiple sectors.

---

[1]*Docking* is a method in molecular modelling which predicts the preferred orientation of one molecule to a second when bound to each other to form a stable complex.

Table 1.2 Classification framework on the basis of biotechnology applications

Very low use	Low/limited research and field application of low-tech only
Low use	Moderate application of low-tech and low/limited research (mostly applied) and mainly on low-tech
Medium use	Significant applications of low-tech, moderate level of research on these technologies and low/limited research (mostly applied) on medium-tech
High use	Low-tech widely applied and there are moderate levels of research on these technologies; there are some low levels of applications of medium-tech; the level of research on medium-tech is moderate
Very high use	High levels of both research and applications on low-tech, medium levels of both research and applications of medium-tech and low or emerging levels of research and applications of high-tech

References

African Union (2020) Africa trade statistics. Yearbook 2020. African union, Addis Ababa. Available via AU. https://au.int/sites/default/files/documents/39607-doc-af-trade_yearbook2020_v4_comp-compresse_1.pdf. Accessed 29 June 2021

AGRA (2017) Africa agriculture status report: the business of smallholder agriculture in sub-Saharan Africa (Issue 5). Nairobi: Alliance for a green revolution in Africa (AGRA). Available at: Final-AASR-2017-Aug-28.pdf (agra.org). Accessed 29 June 2021

AUC (2014) The Malabo declaration on accelerated agricultural growth and transformation for shared prosperity and improved livelihoods, Malabo, Equatorial Guinea, June 26–27, 2014, African Union Commission (AUC), Addis Ababa, Ethiopia. Available at: https://www.resakss.org/sites/default/files/Malabo%20Declaration%20on%20Agriculture_2014_11%2026-.pdf. Accessed 5 July 2021

Bailey R, Willoughby R, Grzywacz D (2014) On trial: agricultural biotechnology in Africa. Energy, environment and resources. Chatham House, London

Bjornlund V, Bjornlund H, Van Rooyen AF (2020) Why agricultural production in sub-Saharan Africa remains low compared to the rest of the world – a historical perspective. Int J Water Resour Dev 36:S20–S53. https://doi.org/10.1080/07900627.2020.1739512

Brink JA, Woodward BR, DaSilva EJ (1998) Plant biotechnology: a tool for development in Africa. Electron Biotechnol 1(3):142–151

CBD (1992) Parties to the Cartagena protocol and its supplementary protocol on liability and redress. Status of ratification and entry into force. Convention on Biological Diversity, Montreal. Available at: https://bch.cbd.int/protocol/parties/. Accessed 1 July 2021

Fan SM, Johnson A, Saurkar M et al (2008) Investing in African agriculture to halve poverty by 2015. IFPRI discussion paper 00751. International Food Policy Research Institute, Washington, DC

FAO (2001) Glossary of biotechnology for food and agriculture – a revised and augmented edition of the glossary of biotechnology and genetic engineering. The Food and Agriculture Organizations of the United Nations (FAO), Rome. Available via FAO. http://www.fao.org/3/Y2775E/y2775e00.htm#Contents. Accessed July 5 2021

FAO, IFAD, UNICEF et al (2020) The state of food security and nutrition in the world 2020. Transforming food systems for affordable healthy diets. FAO, Rome. Available at: https://doi.org/10.4060/ca9692en

Garrity D (2002) Andreas Deininger (ed) attacking the tragedy of hunger and desperate poverty through pro-poor agricultural science. Challenges to organic farming and sustainable land use in the tropics and sub-tropics: international research on food security, natural resource management and rural development, Deutscher Tropentag 2002, Kassel University Press, Kassel

Lipton M (2001) Reviving global poverty reduction: what role for genetically modified plants? J Int Dev 13(7):823–846. https://doi.org/10.1002/jid.845

USDA (2016) Biotechnology frequently asked questions (FAQs). In: U.S. Department of Agriculture. Available via USDA: https://www.usda.gov/topics/biotechnology/biotechnology-fre quently-asked-questions-faqs. Accessed 29 June 2021

WMO (2020) State of the climate in Africa 2019. WMO-No.1253. World Meteorological Organization, Geneva Available at: https://library.wmo.int/doc_num.php?explnum_id=10421. Accessed 1 July 2021

World Bank (2021) Agriculture, forestry, and fishing, value added (% of GDP). World Bank national accounts data, and OECD national accounts data files. World Bank Group, Washington. https://data.worldbank.org/indicator/NV.AGR.TOTL.ZS. Accessed 1 July 2021

The State of Capacities for Agricultural Biotechnology Applications in Crop and Livestock Sectors

2

J. E. O. Rege, Joel W. Ochieng, and Dionysious Kiambi

Abstract

Sub-Saharan African countries were categorized with regard to their capacity for agricultural biotechnology research and application, including human capacities, institutions and facilities, operational budgets and existence of facilitating networks. For the crop and livestock sectors combined, no countries were categorized as having 'very high' capacity; only South Africa was categorized as 'high'; and three countries, Ethiopia, Kenya and Nigeria, were categorized as 'medium'. All other SSA countries were categorized as having 'low' or 'very low' capacity. Capacity was generally higher for crops than for livestock. For the crop sector, South Africa was categorized as 'very high'; Cameroon, Ethiopia, Ghana, Kenya, Nigeria, Sudan, Tanzania, Uganda and Zimbabwe as 'high'; and Botswana, Burkina Faso, Côte d'Ivoire, DRC, Madagascar, Malawi, Mali, Mozambique, Namibia, Rwanda, Senegal, Eswatini and Zambia as 'medium' capacity. For livestock, South Africa was 'high' and Kenya and Nigeria 'medium'. Strong links to international partners, such as hosting a CGIAR centre, tended to be associated with higher capacity.

J. E. O. Rege (✉)
Emerge Centre for Innovations-Africa, Nairobi, Kenya
e-mail: ed.rege@emerge-africa.org

J. W. Ochieng
Agricultural Biotechnology Programme, University of Nairobi, Nairobi, Kenya

D. Kiambi
African Biodiversity Conservation and Innovations Centre, Nairobi, Kenya

2.1   General Classification Framework for Biotechnology Capacities

Key dimensions of capacity used in the analysis in this chapter for both crop and livestock sectors are human capacities, institutions and facilities (including labs, equipment and operational budget) and networks and networking. Capacity categories applied were 'very low', 'low', 'medium', 'high' and 'very high'. The overall framework used is summarized in Table 2.1.

2.2   Crops

2.2.1   Human Resources

No comprehensive study to establish the status of human capacity in terms of full-time equivalent staffing (FTEs) engaged in biotechnology research and development at different levels has been conducted in the recent past in SSA. However, Agricultural Science and Technology Indicators (ASTI) (Box 2.1) provides agricultural research and development statistics, including data on FTEs engaged in agricultural research and development (agricultural R&D) although the data is not disaggregated into biotechnology and/or other fields (IFPRI no date). In this chapter, estimates of FTEs engaged in biotechnology have been made based on the ASTI data supplemented with additional information from desk studies.

Box 2.1 The Agricultural Science and Technology Indicators (ASTI)
Quantitative information is fundamental to understanding the contribution of agricultural science and technology to agricultural growth. Indicators derived from such information allow the performance, inputs and outcomes of agricultural S&T systems to be measured, monitored and benchmarked. These indicators assist science and technology stakeholders in formulating policy, setting priorities and undertaking strategic planning, monitoring and evaluation. They also provide information to governments, policy research institutes, universities and private-sector organizations involved in public debate on the state of agricultural science and technology at national, regional and international levels.

The Agricultural Science and Technology Indicators (ASTI) compiles, analyses and publicizes data on institutional developments, investments and capacity trends in agricultural R&D in low- and middle-income countries worldwide. ASTI has published a set of country briefs and regional synthesis reports that describe general human and financial capacity trends in agricultural R&D at national, regional and global levels. ASTI comprises a network of national, regional and international agricultural R&D agencies and is hosted

(continued)

Table 2.1 Classification framework of countries on the basis of capacities for biotechnology applications

Capacity category	Capacity measures			
	Staffing	Labs/ equipment	Financial resources	[a]R&D networks—public/private; national/ international
Very low	Very low numbers and low training levels	No functional labs	Hardly any budget allocated and no/limited capacity to raise external funding	Hardly any—because of relative inactivity
Low	Low numbers and training levels	Some basic lab facilities but highly inadequate equipment	Limited (occasional) budget for small projects; very limited ability to raise external funding	Limited: R&D work tends to be limited in scope and done by individual institutions in rather sporadic and isolated manner
Medium	Although more is needed, there is significant activity driven by a critical mass of staff	Labs available but equipment is limited	Regular but inadequate budget allocation for biotech research; institutions raise limited extra resources	Opportunistic networks linked to funding and not linked to clear long-term goals and plans
High	Critical mass of personnel for low- and medium-tech research and applications	Well-equipped labs (including BSL-2)	Regular budget allocation that is generally adequate, supplemented by reasonable levels of extra grants for special projects	R&D institutions have considerable strategic networks in and out of the country
Very high	Critical mass of personnel for low- and medium-tech research and applications; some capacity research on high-tech applications	Very well-equipped labs, including BSL-2 and 3	Regular budget allocation; public and private institutions consistently able to raise extra resources for projects	Extensive and dynamic set of R&D networks—in and outside the country—that are strategic and linked to clearly articulated goals

[a]R&D networks: defined here to include networks for the conduct of research and those which facilitate adaptation for use, industry adoption and scaling

Box 2.1 (continued)

and facilitated by the International Food Policy Research Institute (IFPRI). ASTI is currently funded by the Bill and Melinda Gates Foundation.

ASTI's work primarily focuses on (a) initiating institutional surveys in sub-Saharan Africa, Latin America and the Caribbean, the Asia Pacific and the Middle East and North Africa; (b) developing and maintaining a comprehensive, user-friendly website offering access to primary data sources; and (c) building a network of national, regional and international partners to facilitate data collection efforts and the dissemination of outputs.

Since 2009, ASTI launched a web application that allows its users to display different ASTI indicators by country and enables graphical representations of two indicators against each other. Current indicators include

- Trend data on agricultural scientist numbers and total investments in agricultural research by the government, higher education and non-profit sectors of developing countries
- Additional short-term or yearly data on numbers of scientists by degree status and gender, support-staff numbers, funding sources, categories of spending (salaries, operating costs and capital investments) and research focus by agricultural subsector and theme, as well as by sectors, e.g. crop and livestock

ASTI publications include regional and global analyses of agricultural R&D investments and country briefs and fact sheets presenting national data.

Figure 2.1 presents total crop agricultural R&D FTEs and biotech FTE content of this (only MSc and PhD levels) for countries with data. The top 10 SSA countries with highest FTEs are Ethiopia, Nigeria, Tanzania, Kenya, DRC, Ghana, Uganda, Burkina Faso and Côte d'Ivoire, plus South Africa (not shown). The majority of these countries are also among the top in biotechnology applications. Figure 2.1 presents total crop agricultural R&D FTEs as well as crop biotech FTEs expressed per million inhabitants of the countries. Mauritius, Cape Verde and Botswana emerge as countries with among the highest 'intensity' of personnel. Ethiopia follows these countries closely on the basis of this metric.

According to a study carried out by FARA (2011), investments in human resources for biotechnology research appeared to be on an upward trend. However, human capital development appeared to be a critical constraint in many countries surveyed, for example, where there were fewer than 15 biotechnology experts in the entire country. Many of the countries were dependent on experts from ancillary disciplines to complement the human resources needed for biotechnology research. Most of the countries have at least some crop breeders or geneticists. Genetic engineering tends to be the area with the lowest capacity. This has implications also for capacity of countries to develop policies and regulations that form a critical

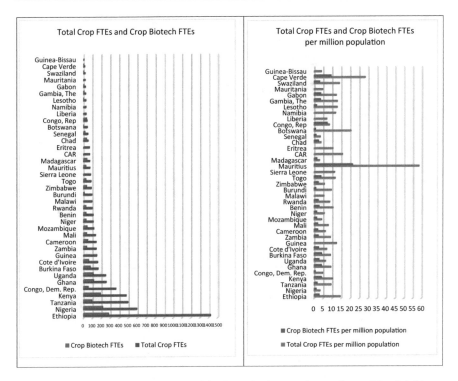

Fig. 2.1 Crop and crop-biotech personnel (FTEs)—absolute numbers and per million inhabitants

part of the enabling environment. For this reason, while FTEs expressed per million inhabitants sound like an appealing metric, it is important that emphasis is put on ensuring a critical mass of personnel of different disciplines and skills. Also important is the fact that, based on the ASTI data, most of the FTEs relevant for biotech in these countries are employees of public agricultural research institutes and universities—institutions historically constrained by operational funds.

Strengthening research capacity and development of expertise in molecular biology, biochemistry, genomics, plant breeding, bioinformatics and policymaking are largely seen as priorities for the effective application of home-grown genetic engineering in SSA agriculture (FARA 2011; Alhassan 2001). Capacity in development and deployment of biotechnology also varies from one type of technology to another with the low-tech applications, such as tissue culture, having a higher number of FTEs in most SSA countries.

In the West and Central Africa subregion, several national research centres have the institutional frameworks to undertake tissue culture research, but the levels of scientific expertise are generally low. Relative intensities of biotechnology capacities also vary widely among countries.

Table 2.2 shows number of plant biotechnologists and breeders per 100,000 hectares of arable land and per million inhabitants for 23 African countries; comparable data was not available for South Africa. Kenya, Ethiopia, Ghana, Zimbabwe

Table 2.2 Relative intensities of biotechnology capacity availability

| | Human resource availability indicator | | | |
| | Per 100,000 hectares of arable land | | Per million inhabitants | |
Country	Plant biotechnologists	Plant breeders	Plant biotechnologists	Plant breeders
Angola	0.1	0.3	0.1	0.7
Benin	0.1	0.8	0.4	2.5
Botswana	0.5	1.1	1.1	2.3
Burkina Faso	0.1	0.9	0.5	3.0
Cameroon	0.3	0.6	1.1	2.3
Côte d'Ivoire	0.6	1.3	1.1	2.4
Eritrea	0.2	7.6	0.3	10.3
Ethiopia	0.3	3.8	2.5	6.0
Gabon	2.8	2.5	7.5	6.7
Ghana	0.6	1.1	2.5	3.1
Kenya	1.4	3.3	3.2	4.0
Malawi	–	1.9	–	3.4
Mali	0.1	0.9	0.5	3.9
Mozambique	–	0.7	–	1.5
Namibia	0.1	2.3	0.5	10.0
Nigeria	0.2	0.5	0.4	1.1
Senegal	0.1	0.4	0.2	0.9
Sierra Leone	0.6	1.1	0.6	1.2
Sudan	0.1	0.3	0.5	1.7
Togo	0.1	0.4	0.4	2.0
Uganda	0.1	0.4	0.2	0.8
Zambia	0.0	0.5	0.1	2.4
Zimbabwe	0.4	1.3	1.0	3.2

Notes: dash = data not available. In this table, Sudan refers to the former Sudan, which is now two independent nations, Sudan and South Sudan
Source: Chambers et al. (2014), plus additional data from ASTI and various other sources from literature

and Gabon were in the top 10 SSA countries ranked on the basis of both plant biotechnologists and plant breeders per 100,000 hectares of arable land and plant biotechnologists and plant breeders per one million inhabitants (Table 2.3). This analysis was instrumental in emphasizing that, when comparing countries, it is important to use metrics that take account of, or adjust for, such important measures as human population and size of the country, and especially available crop and grazing land. Thus, relatively small countries, such as Eritrea and Gabon, may be making good progress in developing capacities required to facilitate deployment of biotechnologies at levels commensurate with the size of the countries' agriculture.

Generally, the ranking of countries in terms of human resources per unit of arable land or per million inhabitants was similar for biotechnologists and plant breeders, although there were exceptions. Human resources (or other investment indicators)

Table 2.3 Ranking of top 10 SSA countries on the basis of human resource availability indicator[a]

Plant biotechnologists per 100,000 ha of arable land	Plant breeders per 100,000 ha of arable land	Plant biotechnologists per million inhabitants	Plant breeders per million inhabitants
1. Gabon	1. Eritrea	1. Gabon	1. Eritrea
2. Kenya	2. Ethiopia	2. Kenya	2. Namibia
3. Sierra Leone	3. Kenya	3. Ghana	3. Gabon
4. Ghana	4. Gabon	4. Ethiopia	4. Ethiopia
5. Cote d'Ivoire	5. Namibia	5. Cote d'Ivoire	5. Kenya
6. Botswana	6. Malawi	6. Cameroon	6. Mali
7. Zimbabwe	7. Zimbabwe	7. Botswana	7. Malawi
8. Ethiopia	8. Cote d'Ivoire	8. Zimbabwe	8. Zimbabwe
9. Cameroon	9. Sierra Leone	9. Sierra Leone	9. Ghana
10. Nigeria	10. Ghana	10. Sudan	10. Burkina Faso

[a]Data on South Africa was not available for this analysis

expressed per 100,000 hectares of arable land or per million inhabitants represent more meaningful metrics for comparing countries than do absolute numbers.

2.2.2 Institutions and Facilities

Most SSA countries have at least some infrastructure and facilities for tissue culture. A study by ASARECA in 2013 revealed that there had been investment made by countries to promote tissue culture application within the eastern and Central Africa (ECA) subregion and that tissue culture infrastructure had progressively increased in both public and private sector over a 10-year period (Masiga et al. 2013). However, most of the institutions do not have a critical mass of human resource, laboratory infrastructure and adequate financial resources to realize the potential of tissue culture in the ECA region. In addition, countries differ significantly with regard to quality of tissue culture infrastructure and involvement of the private sector. For instance, in Kenya eight institutions were directly involved, while in Burundi there were only four institutions. In Rwanda, no private tissue culture laboratory exists, while in Kenya, at least three private laboratories are available, and there are several non-governmental organizations (NGOs) and universities actively involved in tissue culture training and/or awareness.

Key constraints affecting tissue culture research and applications include high operating costs, inadequate facilities for virus indexing and certification, misconceptions about tissue culture products, weak linkages among tissue culture stakeholders, limited local suppliers of tissue culture reagents and inadequate national and/or regional policy frameworks to support private-sector involvement in tissue culture (Masiga et al. 2013).

A number of countries have infrastructural capacities for low-level genomics and molecular marker technologies (Kiome 2015). In the ECA, Ethiopia, Kenya, Sudan and Uganda have the most advanced biotechnology facilities. For instance, in Kenya there are at least 12 public institutions (national agricultural research institutes (NARIs) and universities) with good biotechnology research facilities. This includes two biosafety level-1 (BSL-1) and two BSL-2 facilities. In Uganda, most of the biotechnology facilities are in the National Agricultural Research Organisation (NARO) research stations and Makerere University, with the main facilities being at Kawanda Agricultural Research Institute which has a BSL-1 facility. Sudan has 14 fully functional laboratories operated by the Agricultural Research Corporation (ARC) and universities. These laboratories, including two BSL-2 labs, are well equipped for bio-fertilizer, tissue culture, molecular markers, genomics and genetic engineering work. The ARC's Biotechnology and Biosafety Centre in Khartoum has the capacity for genetic engineering, genetically modified organism (GMO) detection, genomics and national-level biosafety clearing mechanism (Ali 2009). Ethiopia has substantial biotechnology infrastructural capacities in the National Agricultural Biotechnology Research Centre, the Ethiopian Biodiversity Institute, public universities and various research centres of the Ethiopia Institute of Agricultural Research (Abraham 2009). Most of the other countries in the ECA subregion only have capacities to conduct low- to medium-level biotechnology with some, like Djibouti and Somalia, having close to none. Despite its otherwise low overall biotech capacity, Gabon has one of only two BSL-4 facilities in Africa, but these were established and primarily utilized for work on the Ebola virus and are of little or no relevance to agricultural biotechnology.

South Africa is by far the most advanced country in SSA for biotechnology research facilities and infrastructure. The country has three BSL-2 and one BSL-3 facilities, and at the Centre for Emerging Zoonotic and Parasitic Diseases, part of the National Health Laboratory Service, is the only suit BSL-4 containment facility in Africa. The latter provides the highest level of safety allowing work on agents that can easily be aerosol-transmitted within the laboratory and can cause severe to fatal disease in humans for which there are no available vaccines or treatments. The BSL-2 and BSL-3 facilities are in the Agricultural Research Centre (ARC) and universities. The ARC biotechnology platform operates genotyping research facilities, including three next generation Illumina sequencers, a HiSeq2500 and two MiSeq systems.

In Malawi there are four molecular biology laboratories, seven tissue culture/ transformation laboratories and one BSL-2 laboratory at Lilongwe University of Agriculture and Natural Resources (LUANAR). Most of these facilities are in the newly established biotechnology research and development laboratory at LUANAR or in government agricultural research stations. Other countries in the southern Africa subregion with substantial biotechnology infrastructural capacities include Botswana, Namibia, Zambia and Zimbabwe (Olembo et al. 2010).

In West Africa, Ghana has established a national biotechnology research and development authority complete with offices and requisite laboratories under the Crops Research Institute of the Council for Scientific and Industrial Research (CSIR). Within the country, there are nine molecular biology laboratories and nine tissue culture/transformation laboratories in various research centres under CSIR and local universities, including one BSL-1 and one BSL-2 (Quain and Asibuo 2009). Nigeria has a well-equipped biotechnology laboratory at the Institute of Agricultural Research. The country also has several other biotechnology laboratories in its various NARIs and numerous universities, including one BSL-1 and BSL-2. Cameroon has a biotechnology research institute at the Institute for Research on Agricultural Development, but it is not staffed and does not have equipment. Burkina Faso and Côte d'Ivoire have reasonably good biotechnology infrastructural capacities. Most of the other countries in the West African subregion only have minimal infrastructural capacities to conduct low- to medium-level biotechnology R&D.

Overall, except for South Africa and the facilities at the International Livestock Research Institute-Biosciences eastern and central Africa (ILRI-BecA) Hub in Kenya, no other SSA country has significant equipment for genomics research, such as sequencers or bioinformatics platforms. Table 2.4 presents a summary of transgenic laboratory research capacities in some SSA countries.

Availability and maintenance of infrastructure needed for genetic engineering remains one of the most limiting factors for most SSA countries. Many NARS have fragile crop research programmes that are often dependent on a handful of scientists. Faced with human, financial and infrastructural constraints, many such programmes are unable to implement sustained initiatives beyond pilot scales (Chambers et al. 2014; Kiome 2015). In addition, many laboratories do not function optimally because of inadequate power supplies and frequent breakdowns in their research equipment. The present lack of engineers trained to service the sophisticated equipment required in genetic engineering research compels frequent recourse to manufacturing firms abroad for essential repairs. It is therefore necessary that in parallel with the development of genetic engineering research infrastructure, African countries should also build an infrastructure and capacity for equipment repair and maintenance (Alhassan 2001; FARA 2011; Chambers et al. 2014; Kiome 2015).

2.2.3 Networks and Networking

Recognition of the need for capacity and infrastructure to speed up development and uptake of biotechnologies—to address Africa's major agricultural challenges—has led to various initiatives. While most of these initiatives and networks associated with them are also relevant for livestock, forestry and aquaculture, they are presented in detail here because of their primary focus on crops.

A key one is the African Biosciences Initiative (ABI) operating under Africa's Science and Technology Consolidated Plan of Action of the African Union Commission (AUC) and the New Partnership for Africa's Development (NEPAD).

Table 2.4 Summary of facilities for transgenic research in selected SSA countries

Country	Institution	Number of laboratories for				Number of BSL labs		
		Mol.	Seq.	TF	BIF	BSL-1	BSL-2	BSL-3
Benin	Institute of Environment and Agricultural Research	1	0	0	0	1	0	0
Burkina Faso	Institut National des Recherches Agricoles du Bénin	1	0	0	0	1	1	0
Burundi	Institut des Sciences Agronomiques du Burundi	1	0	0	0	0	0	0
Cameroon	National Biotechnology Research Institute (University of Yaoundé)	3	0	0	0	0	0	0
Ethiopia	Ethiopian Institute of Agricultural Research, universities	2	0	0	0	0	0	0
Ghana	NARS and universities	1	0	1	0	1	1	0
Kenya	Kenya Agricultural and Livestock Research Organization, universities	2	0	1	0	2	2	0
Malawi	Department of Agricultural Research Services and Universities	1	0	0	0	1	0	0
Nigeria	NARS and universities	3	0	1	0	1	1	0
South Africa	Biotechnology Research Institute, ARC	2	1	2	1	1	3	2
Uganda	NARO, universities	2	0	1	0	1	1	0
Zimbabwe	Department of Agricultural Research and universities	1	0	0	0	1	0	0

Source: Kiome (2015) and additional information from desk review
[a]Note: Mol = molecular; Seq = sequencing; TF = transformation; BIF = bioinformatics platform; use of these facilities are not confined to transgenic work only

Under one of its programme clusters, Biodiversity, Biotechnology and Indigenous Knowledge, the ABI committed to facilitating the establishment of regional networks of laboratories with state-of-the-art facilities. They were referred to as Centres of Excellence. Four such networks were to be established, starting in the mid-2000s: BecA, Southern Africa Network for Biosciences (SANBio), West Africa Biosciences Network (WABNet) and North Africa Biosciences Network (NABNet).

Each network was supposed to implement regional flagship research programmes and capacity building activities in different disciplines that reflected identified priorities and comparative advantages based on nucleus of work and capacities then present in the hosting institution: BecA focused on agricultural sciences broadly, SANBio on human health and nutrition, WABNet on crop-based agriculture and food security and NABNet on agriculture/food security and human health.

The Biosciences eastern and central Africa (BecA) (Box 2.2) was established at ILRI in Nairobi in 2003. It stands out in comparison with the others as an example of a shared facility which has successfully provided support to eastern and Central African countries and beyond to develop and apply bioscience research and expertise in agriculture. While the majority of those who have collaborated with and used the BecA Hub to date are from crops and livestock R&D disciplines, the platform is designed to be generic and is relevant for forestry as well as aquaculture. Indeed, the Hub has contributed to agroforestry as well as aquaculture through research and training. Neither WABNet nor NABNet have been operationalized as envisaged under the ABI. They were not built around any lab infrastructure. Instead, they functioned for a few years as loose networks but never matured to become vibrant and visible. Indeed, they have become moribund with time and have not yet lived up to the original expectation, principally due to resource challenges. The SANBio Hub, located within the Biosciences Unit at the CSIR, South Africa, has had modest success, but its focus is mostly on human health.

In recent years, the World Bank's regional approach to programming through its East African Agricultural Productivity Program (EAAPP), West Africa Agricultural Productivity Program (WAAPP) and Agricultural Productivity Programme for Southern Africa (APPSA) has made some progress in addressing SSA's most acute agricultural research capacity challenges, in rehabilitating research infrastructure and in funding priority research areas. However, these programmes have had a focus on agricultural R&D more broadly, and not biotechnology, and laboratory infrastructure development has been modest, at best.

Box 2.2 Institutional Arrangements to Build Capacity for, and Increase Use of, Agricultural Biotechnology in SSA: The Case of the Biosciences Eastern and Central Africa (BecA)

The Biosciences eastern and central Africa (BecA) Hub aimed to support eastern and Central African countries to develop and apply bioscience research and expertise. The vision of the BecA Hub—hosted by ILRI in Nairobi and initiated in 2003—was to enable African scientists and institutions to become significant technological innovators as well as users through access to state-of-the art biosciences facilities in which high priority, cutting-edge biotechnology research addressing Africa's problems would be conducted.

The BecA Hub core competencies include genomics/metagenomics, bioinformatics, genetic engineering, diagnostics, molecular breeding, vaccine

(continued)

Box 2.2 (continued)

technology/immunology and mycotoxins. Core experts at the Hub consist of staff of ILRI's biotechnology and related programmes as well as scientists and technical staff from a range of partner institutions hosted at the Hub. In 2018/2019, the FTEs at the Hub consisted of 49 core BecA staff, and it awarded 78 Africa Biosciences Challenge Fund (ABCF) fellowships to young researchers in 12 countries and hosted 13 communities-of-practice, while the Bioinformatics Platform trained 130 African scientists. This combination created a melting pot for science and innovation at the Hub—a platform for learning and networking (BecA-ILRI Hub 2020).

BecA had capacity for upwards of 50 scientists annually through short- and long-term (including MSc and PhD) courses. In addition, the Hub hosted several group courses on biotechnology topics every year. By 2018/2019 BecA was well recognized as the major shared agricultural biotechnology research Hub in SSA. Despite its name, BecA Hub users have included scientists (faculty as well as fellows/trainees) from across the continent. The Hub has indeed presented a major opportunity for SSA countries looking for training in various areas of biosciences. However, while the Hub was designed with capacity to accommodate 250 FTE users (including faculty), this potential was never fully harnessed (see graph below), the highest occupancy having been 159 FTEs in 2014. Thus, while there is demand for such facilities by countries, this demand is not supported by allocation of resources, and users of BecA have generally tended to be those sponsored by donors. There has been idle capacity whose full exploitation will need a different funding approach.

The resourcing model envisaged for the Hub has not worked. The establishment of the facilities was funded by the Canadian government as once-off capital investment. Access to facilities was to be funded by NARS and from grants secured by NARS working with their partners. The model for funding operations finally collapsed, and by 2019 ILRI management decided to examine an alternative model driven and funded by CGIAR programmes in Africa, with NARS training being embedded and accommodated only to the extent of alignment to CGIAR priorities, funding and scope of programmes. Discussions on form and content focus are currently (2021) ongoing to transform BecA into what is currently being called 'Biosciences for Africa' (B4A).

(continued)

Box 2.2 (continued)

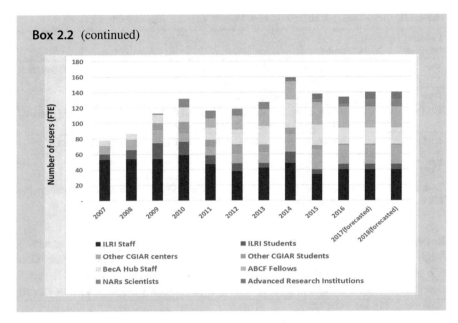

Another World Bank initiative is the Africa Higher Education Centres of Excellence (ACE), which was designed to deliver high-quality and relevant postgraduate education in priority fields, including agriculture, health and other sciences. Its first phase, ACE I, was launched in 2013 for West and Central Africa, and 22 Centres of Excellence across 8 countries in priority sectors—agriculture being one of these— were supported. The aim was to address regional development challenges. Building on the ACE I experience, the second phase, the Eastern and Southern Africa Higher Education Centers of Excellence (ACE II), launched in 2016, aimed to strengthen selected ESA institutions to deliver quality postgraduate education and build collaborative research capacity in five regional priority areas: industry, agriculture, health, education and applied statistics. The development objective of ACE is to support the recipients to promote regional specialization among participating universities in areas that address regional challenges and strengthen the capacities of these universities to deliver quality training and applied research. Between 2014 and 2020, over 14,000 masters and PhD students, 30% of whom were women, were supported in agriculture, health and other sciences.

2.2.4 Classification of Countries

Drawing on considerations of human capacities, institutions and facilities, operational budgets and existence of facilitating networks and networking, SSA countries can be classified according to their overall biotechnology capacity for crops as summarized in Table 2.5. Only South Africa is classified as having 'very high' capacity for biotechnology for crops. Nine countries, Cameroon, Ethiopia, Ghana,

Table 2.5 Classification of countries on the basis of biotechnology capacity for crops

Capacity category	Countries
Very low	Angola, Chad, Central Africa Republic, Congo, Djibouti, Gambia, Equatorial Guinea, Guinea, Guinea Bissau, Liberia, Somalia, South Sudan
Low	Benin, Burundi, Gabon, Eritrea, Lesotho, Niger, Sierra Leone, Togo
Medium	Botswana, Burkina Faso, Côte d'Ivoire, DRC, Eswatini, Madagascar, Malawi, Mali, Mozambique, Namibia, Rwanda, Senegal, Zambia,
High	Cameroon, Ethiopia, Ghana, Kenya, Nigeria, Sudan, Tanzania, Uganda, Zimbabwe
Very high	South Africa

Countries not listed due to lack of sufficient data: Cape Verde, Comoros, Mauritania, Mauritius, Sao Tome and Principe and Seychelles

Kenya, Nigeria, Sudan, Tanzania, Uganda and Zimbabwe, were classified as having 'high' capacity and 13 as having 'medium' capacity (Botswana, Burkina Faso, Côte d'Ivoire, DRC, Madagascar, Malawi, Mali, Mozambique, Namibia, Rwanda, Senegal, Eswatini, Zambia). The remaining 20 countries were classified as having 'low' or 'very low' capacity.

2.3   Livestock

Capacity for research, development and application of biotechnologies in the livestock sector is, even more than for crops, highly variable across countries. The specific dimensions of capacity in this connection and the status in SSA countries are described briefly in the sections below.

2.3.1   Human Resources

Figure 2.2 summarizes the human resources for livestock in public agricultural R&D in SSA countries. The top 10 countries with regard to livestock personnel (FTEs), excluding South Africa, are Ethiopia, Nigeria, Kenya, Uganda, Tanzania, Mozambique, Botswana, Ghana, DRC and Cameroon, in that order. Compared to crop FTEs, there is some change in ranking: Burkina Faso, Côte d'Ivoire and Zambia are no longer in the top 10.

On average across SSA, the crop sector contributes a larger proportion of the agricultural GDP than livestock: for the period 1990–2013, crops accounted for 85% of the total production value in the region. However, the actual contribution of livestock varies between the SSA regions and countries and does not seem to be fairly reflected in the public investments in the sector. The only country where livestock's share of agriculture R&D FTEs (64 FTEs) is higher than crops (45) is Botswana, where human resources dedicated to livestock represents 59% of the total agriculture R&D FTEs. However, even in this case the allocation does not

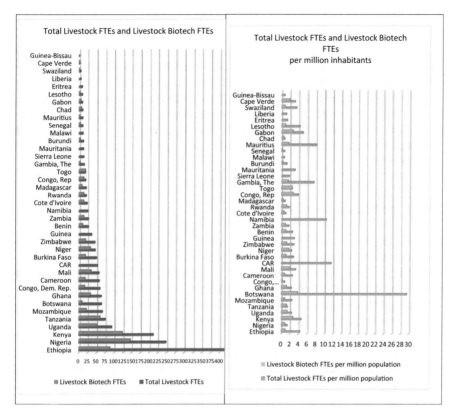

Fig. 2.2 Livestock and livestock-biotech personnel (FTEs)—absolute numbers and per million inhabitants

proportionately reflect the contribution of livestock to agricultural GDP, which is 85%. Another example is Ethiopia where livestock, even excluding the contribution of manure, draught power and transport, accounts for 40% of agricultural GDP, and yet 77% of agriculture R&D FTEs is in the crop sector.

The situation is even worse when one considers the disciplinary content of the FTEs. The disciplines that have major potential to drive biotechnology applications in livestock are those which underpin animal genetics and breeding, vaccines and diagnostics and, to some extent, nutrition. Relative 'biotechnology contents' show significant variation among countries (Fig. 2.2), with Nigeria, Kenya, Ethiopia, Tanzania and Uganda being the top 5, followed by Ghana, Mali and Niger. The biotech content of livestock FTEs for the rest of the countries is much more limited. For biotech FTEs expressed per one million inhabitants, Botswana leads followed by Kenya, DRC, Togo and Gabon. The small island states of Mauritius and Cape Verde follow closely. Like in the case of crop FTEs, this measure (per million inhabitants) shows that many small countries are doing better than they get credit for. Generally, the number of MSc and PhD FTEs that works on livestock biotechnology tend to be closely related to the total agriculture R&D FTEs. That is, those countries with high total livestock FTEs are also countries with medium to high levels of biotech applications. However, there is a much closer relationship of total agriculture R&D FTEs with crop biotech FTEs (focusing on MSc and PhD levels only) than with livestock FTEs. This indicates that increased investments in capacity at graduate level for agriculture R&D tend to favour crops more than livestock and suggests a need for more intentionality in sector-specific biotech capacity development. This is even more critical considering the poor sharing of human and infrastructural capacities between sectors under the predominant institutional structures in most SSA countries.

2.3.2 Institutions and Facilities

South Africa, Kenya and Nigeria, in that order, lead SSA countries in terms of the existence of well-established and strong livestock research institutions. South Africa has a total of 20 universities and colleges of agriculture. The most well-known of these in the field of livestock research and training are Stellenbosch, Fort Hare, Free State and Pretoria. However, some of the other universities also have training programmes on livestock as well as modest engagements in livestock biotechnology research.

Significant mandate and research on livestock in South Africa rests with the ARC, primarily through the Onderstepoort Veterinary Institute (OVI), whose major programmes are on animal vaccines, diagnostics and therapeutics and, in the Animal Production Department, animal genetics, breeding, nutrition, rangelands and other aspects of production.

Kenya has 12 universities with programmes in various aspects of animal biotechnology. The oldest of these are the College of Agriculture and Veterinary Sciences of the University of Nairobi and the Department of Animal Sciences of Egerton University. In addition, biotechnology programmes being undertaken in four other public universities—Kenyatta, Jomo Kenyatta, Moi and Eldoret universities—while focused on crops and other sectors are making contributions to livestock through relatively modest and new initiatives consisting primarily of graduate research projects. Most of these projects are on aspects of molecular characterization of livestock populations. The Kenya Agriculture and Livestock Research Organization (KALRO, formerly KARI) has 6 institutes out of a total of 16 that are dedicated to livestock research plus 2 (Biotechnology Institute and Genetic Resources Institute) with significant livestock biotechnology content. Vaccine production in Kenya is the responsibility of a separate government entity, KEVEVAPI, which currently produces 14 mostly attenuated vaccines. Research on vaccines and diagnostics is carried out by the Veterinary Science Research Institute of KALRO.

Nigeria has 18 national agricultural research institutes, 3 federal colleges of agriculture, 47 faculties of agriculture and 8 faculties of veterinary medicine. Nigeria's National Animal Production Research Institute (NAPRI), which is affiliated to the Ahmadu Bello University in Zaria, is specifically responsible for livestock R&D with a focus on production aspects, and the National Veterinary Research Institute is responsible for animal health R&D. Through a combination of crossbreeding and selection, NAPRI scientists have developed the Shika Brown chicken, and the Federal University of Agriculture, Abeokuta, has developed the FUNAAB-alpha chicken. Both of these are now recognized as distinct breeds within the country. In both cases, the breed development process is ongoing.

At the other end of the spectrum are the majority of countries with hardly any livestock research, let alone biotechnology. These countries do have institutions established and mandated to lead livestock R&D, but they have limited human or infrastructural capacity to design and implement credible research on modern livestock biotechnologies, beyond the lower-end applications of well-established technologies such as crossbreeding or use of applied animal nutrition research, for example, testing animal performance on different feeds. Countries at this low end tend to have no university-level livestock biotechnology research and training and do not have capacity to develop technologies domestically. Some of them have had projects supported by FAO and the International Atomic Energy Agency (IAEA), but these have been short-term projects many of which were not sustained following the end of project funding. So, while these projects have helped expose local scientists to some technologies, these learnings have not yet led to subsequent home-grown programmes. Countries in this category include Angola, Benin, Burundi, Chad, Central Africa Republic, Congo, Djibouti, DRC, Gambia, Eritrea and South Sudan, among others. Some of the countries in this cluster have some capacity in specific areas; for example, Angola has, for many years, been involved in vaccine production which is recognized by the World Organisation for Animal Health (OIE).

Countries in between these extremes—with medium-level institutional capacities—are those which have some basic biotechnology R&D infrastructure in research institutes (NARIs) and universities and which, despite inadequate labs and equipment, have had some successful projects/programmes on one or more of the livestock biotechnologies. These are mostly low-tech, but in some cases also medium-tech. Examples include such countries as Ethiopia, Tanzania, Sudan, Uganda, Ghana, Mali, Botswana and Senegal, all of which can be categorized as medium with regard to livestock R&D institutions. Livestock research, including biotechnology, in these countries is driven by strong NARIs, supported by constituent schools or colleges of veterinary and/or animal sciences in increasing numbers of universities of agriculture in these countries.

The Botswana Vaccine Institute (BVI) produces foot-and-mouth disease (FMD) vaccine, its major focus, as well as vaccines for peste des petits ruminants, contagious bovine pleuropneumonia, anthrax and black quarter. Since 1985, BVI has hosted and administered the OIE Sub-Saharan Africa Regional Reference Laboratory for FMD. In this function, BVI is tasked to provide confirmatory laboratory diagnosis of FMD outbreaks to all AU member states in SSA, particularly countries in southern Africa; bench training on FMD diagnostic techniques (virus isolation and typing, PCR and sequencing, antibody detection by ELISA and virus neutralization test); and harmonization of laboratory techniques and methods. Clearly, despite the fact that Botswana is not engaged in high-level livestock biotechnology research in other areas, BVI places the country at almost the same footing as Kenya and Nigeria in the area of diagnostics. Botswana is also well known for its beef industry which is underpinned by a strong breeding programme which includes a functioning government-driven artificial insemination (AI) field service in beef herds, incorporating crossbreeding and within-breed selection.

Also in this cluster of countries is Ethiopia. Ethiopia has a long history of livestock improvement and conservation programmes focusing on indigenous cattle, sheep, goats and chickens. The Ethiopian AI service, although relatively small in coverage, has a long history having first been established in 1938. In 1984, the present National Artificial Insemination Center (NAIC) was established to coordinate the overall AI operation at national level (GebreMedhin 2005) and has been running, albeit with limited coverage, since then. Unlike in many SSA countries, where the programmes have had to shut down at some point, NAIC has received government support and has never had to close down, even during the difficult economic hardship during the Derg regime (1974–1991).

Ethiopia's National Veterinary Institute produces 16 livestock vaccines, and the diagnostic laboratory system is well equipped and staffed with 14 diagnostic laboratories currently available nationwide. Ethiopia has also been engaged in poultry breeding. Starting in 2008 it has successfully developed an improved local (Horro) chicken strain through within-population selection which has doubled egg production. Of the eight Ethiopian universities, at least three (Haramaya, Awassa and Jimma) have capacity for research on livestock biotechnology, at least on the low-level biotechnologies. While not an agricultural university, Addis Ababa University has been involved in agricultural biotechnology research, mostly crops, but the basic biotechnology labs and scientists have potential to contribute to livestock biotech R&D. Indeed, some faculty members have collaborated with ILRI and other Ethiopian universities on livestock-related biotechnology projects.

Another country in this medium cluster is Mali. Mali's Central Veterinary Laboratory (LCV) in Bamako is a well-established facility. In existence since 1979, LCV has activities in vaccine production, development and adaptation of diagnostics tools and training of animal health personnel. LCV is an example of a national institution that has flourished through deliberate focus on networking. Through sustained and evolving partnerships with international agencies around the world, including USAID, IAEA and FAO, and international research centres such as CIRAD-EMVT (France), ILRI, CIRDES (based in Bobo, Burkina Faso), the LCV has become one of the leading labs in diagnosis and research in animal diseases and production of animal vaccines in West Africa. LCV makes a range of animal vaccines for use within Mali as well as exporting to Burkina Faso, Togo, Benin, Guinea Conakry, Mauritania, Côte d'Ivoire and DRC. In 2009, LCV infrastructural capacity got a significant boost when it received new equipment and facilities from a 6.9 million euros investment through the GALVmed-led component of an AU-IBAR-led initiative, the VACNADA (Vaccines for Control of Neglected Animal Diseases in Africa) project, being one of the eight labs across Africa that have benefited from this initiative. The other countries whose labs benefited from

VACNADA support through this project were Botswana, Cameroon, DRC, Ethiopia, Ghana, Kenya and Senegal (GALVmed 2012).

Besides South Africa and Kenya, both of which have well-developed AI programmes, Mali is among the few SSA countries (others being Nigeria, Ethiopia, Uganda, Ghana, Botswana, Malawi, Senegal and Sudan) which have taken the AI technology to the field, mostly for upgrading indigenous cattle but also to support pure breeding programmes in limited numbers of commercial farms. In 2014, Mali launched a major new AI initiative supported by a donation of 125,000 doses of semen by the Mohamed VI Foundation of Morocco. The new Mali-Morocco cooperation in this field includes training and retooling of Malian inseminators. The programme was expected to lead to the insemination of 10,000 cows per year for 5 years and presents a major opportunity for Mali to be a significant user of AI.

2.3.3 Networks and Networking

Networks facilitating biotechnology applications consist of various bilateral partnership arrangements between SSA countries, especially involving cross-border movements of semen and vaccines, as well as collaborations with regional and international institutions. For example, similar to the regional presence established by Mali's Central Veterinary Laboratory (CVL) in the animal health domain (see Sect. 2.3.2), the Agricultural Research Institute for Development in Cameroon supports AI services in the Central Africa Republic, while semen produced in Kenya is used in multiple countries in East Africa, and South Africa's semen is exported to countries in southern and eastern Africa. Vaccine produced by Nigeria, Kenya, Botswana and South Africa, and a few other countries, are exported to neighbouring countries and in some cases beyond.

Collaborations involving regional and international institutions play a significant role in facilitating applications of, and capacities for, biotechnologies. ILRI is the major CGIAR centre working on livestock in Africa. It has a history of significant capacity development in all subregions of SSA, but with eastern Africa, especially Kenya and Ethiopia, which host major facilities of ILRI, being the major beneficiaries. ILRI's livestock biotechnology capacity building has been primarily in genetics and breeding, breed characterization including disease resistance/tolerance as well as vaccine technology. The International Center for Agriculture Research in the Dry Areas (ICARDA) has also had some impact on capacity development in applied breeding and breed characterization, especially around the Horn of Africa with Ethiopia being a major beneficiary.

2.3.4 Classification of Countries

In Table 2.6, countries are classified according to their overall biotechnology capacities for livestock. None were classified as 'very high' capacity; only South Africa was classified as 'high'; and, just two countries, Kenya and Nigeria,

Table 2.6 Classification of countries on the basis of biotechnology capacities for livestock

Capacity category	Countries
Very low	Angola, Benin, Burundi, Chad, Central African Republic, Republic of Congo, Djibouti, DRC, Equatorial Guinea, Eritrea, Eswatini, Gabon, Gambia, Guinea, Guinea Bissau, Lesotho, Liberia, Niger, Sierra Leone, Togo, Somalia, South Sudan,
Low	Botswana, Burkina Faso, Cameroon, Côte d'Ivoire, Ethiopia, Ghana, Madagascar, Namibia, Malawi, Mali, Mozambique, Rwanda, Senegal, Sudan, Tanzania, Uganda, Zambia, Zimbabwe
Medium	Kenya, Nigeria
High	South Africa
Very high	None

Countries not listed due to lack of sufficient data: Cape Verde, Comoros, Mauritania, Mauritius, Sao Tome and Principe and Seychelles

were classified as 'medium'. All other SSA countries, the vast majority, were classified as 'low' or 'very low' capacity. This underscores the status of livestock R&D generally, and livestock biotechnology specifically. Examined against the status of crops capacities (in Sect. 2.2), this classification reveals the disconnect in some countries between the livestock subsector's contribution to agriculture GDP and investments being made in human capacity and infrastructure to support livestock R&D, including biotechnology.

References

Abraham A (2009) Agricultural biotechnology research and development in Ethiopia. Afr J Biotechnol 8(25):7196–7204

Alhassan WS (2001) The status of agricultural biotechnology in selected West and Central African countries. International Institute of Tropical Agriculture, Ibadan, Nigeria

Ali AM (2009) Status of biotechnology in Sudan. In: Madkour M (ed) Status and options for regional GMOs detection platform. A benchmark for the region. FAO, Rome

BecA-ILRI Hub (2020) BecA-ILRI hub 2018–2019 Biennial report. Biosciences eastern and central Africa-International Livestock Research Institute Hub (BecA-ILRI), Hub, Nairobi

Chambers JA, Zambrano P, Falck-Zepeda J et al (2014) GM agricultural technologies for Africa. International Food Policy Research Institute and Africa Development Bank, Washington and Abidjan

FARA (Forum for Agricultural Research in Africa) (2011) Status of biotechnology and biosafety in sub-Saharan Africa: a FARA 2009 study report. FARA Secretariat, Accra

GALVmed (2012) GALVmed newsletter VACNADA special edition. GALVmed, Edinburgh. Available via GALVmed. https://www.galvmed.org/wp-content/uploads/2016/07/vacnada-newsletter.pdf. Accessed 1 July 2021

GebreMedhin D (2005) All in one: a practical guide to dairy farming. Agri-Service Ethiopia Printing Unit, Addis Ababa

Kiome R (2015) A strategic framework for transgenic research and product development in Africa: Report of a CGIAR study. ILRI, Nairobi

Masiga CW, Mneney E, Wachira F et al (2013) Situational analysis of the current state of tissue culture application in the Eastern and Central Africa Region. Association for Strengthening Agricultural Research In East and Central Africa (ASARECA), Entebbe

Olembo N, M'mboyi F, Oyugi K et al (2010) Status of crop biotechnology in sub-Saharan Africa. African Biotechnology Stakeholders Forum, Nairobi

Quain MD, Asibuo JY (2009) Biotechnology for agriculture enhancement in Ghana: the challenges and opportunities. Asian Biotechnol Dev Rev 11(3):49–61

The State of the Enabling Environment for Agricultural Biotechnology Applications in Crop and Livestock Sectors

3

J. E. O. Rege, Dionysious Kiambi, and Joel W. Ochieng

Abstract

Sub-Saharan African countries were categorized with regard to the state of the enabling environment for agricultural biotechnology in the crop and livestock sectors. This included consideration of public awareness, participation and acceptance; existence of a national biosafety framework; public policy and political goodwill; and the financial and investment environment. South Africa was the only country whose enabling environment was ranked as being 'very strong' for both crops and livestock, while Kenya and Nigeria were in the 'strong' enabling environment category for both crops and livestock. Sudan and Ethiopia were in the 'strong' category for crops but 'medium' category for livestock. Botswana, Ghana, Malawi, Mali, Namibia, Tanzania, Uganda, Zambia and Zimbabwe were in 'medium' category for both crops and livestock. Malawi was 'medium' for crops but 'weak' for livestock. The remaining countries fell in 'weak' and/or 'very weak' for both crops and livestock. Key elements of the enabling environment were driven by actions in the crop sector, especially the pull factor with regard to creating conditions for the use of genetically modified crops. All but two countries, Equatorial Guinea and South Sudan, have ratified or complied with accession requirements for biosafety regulatory systems specified under the Cartagena Protocol on Biosafety.

J. E. O. Rege (✉)
Emerge Centre for Innovations-Africa, Nairobi, Kenya
e-mail: ed.rege@emerge-africa.org

D. Kiambi
African Biodiversity Conservation and Innovations Centre, Nairobi, Kenya

J. W. Ochieng
Agricultural Biotechnology Programme, University of Nairobi, Nairobi, Kenya

3.1   General Framework for Classification of Enabling Environment

The extent of application of any technology by end users is frequently modelled by diffusion theory whose parameters are influenced by several factors, here referred to collectively as the 'enabling factors'. In this chapter, we consider four interacting broad groups of factors: public awareness, participation, and acceptance; national biosafety frameworks; public policy and political goodwill; and financial and investment environment. These factors influence the endogenous development of biotechnologies and/or adoption and use of exogenous biotechnologies by the end users. While factors such as policy and biosafety are major constraints to genetic engineering and other high-tech applications, other factors such as public awareness, acceptance and availability of investments are critical across the board, including for low- and medium-tech applications. The five enabling environment categories used to classify countries are very weak, weak, medium, strong and very strong. The overall framework is summarized in Table 3.1.

3.2   Crops

3.2.1   Public Awareness and Political Support

Despite an increasing body of scientific and economic evidence that shows the overall benefits of crop biotechnology, the polarized debate continues at all levels of African society. The biotechnology debate centres around myriad issues related not only to safety but also to concerns about food security, national sovereignty, farmers' rights, social justice and poverty reduction. The continued back-and-forth nature of the conversation has led to technological paralysis and an inability to make informed decisions among policymakers and members of the public across most of the African continent.

The continuing controversy has deprived African farmers of the ability to make informed and independent choices about which technologies to employ on their farms. Public acceptance and participation seem to be a critical factor in determining the pace at which the enabling environment, including formulation of policies and biosafety frameworks, can be positively changed. To improve the biotechnology enabling environment, different interventions have been made at global, regional, subregional and national levels targeting both policymakers and the general public. Box 3.1 summarizes a case in the Southern Africa Development Community (SADC) region which illustrates this point.

Table 3.1 Classification framework of countries on the basis of enabling environment for biotechnology applications

Status of enabling environment	Existence of policy on biotech application	Operationalization of biotech policy	Existence of Biosafety Framework (Act, Regulation, Agency/Body)	Operationalization of Biosafety Framework	Public and private sector investments including (PPP) in agricultural biotechnology
Very weak	No	Not applicable	No	Not applicable	Very low
Weak	Yes or no	If policy exists, it is not operationalized	In preparation or draft available	Yes, but not operationalized	Low
Medium	Yes	In place and operational—mainly as a reference document in planning—with limited stakeholder awareness	Yes	In place and operationalized—but usage limited (e.g. to research)	Medium
Strong	Yes	In place and operational—used consistently in planning and execution of projects, etc.	Yes	In place and operationalized, actively used (to guide research and commercial or field applications)	High
Very strong	Yes	In place and operational—consistently guides planning, allocation of resources and implementation of programmes	Yes	In place and operationalized; very actively used (research and commercial/field applications)	Very high

Box 3.1 Biotechnology Outreach and Information Transfer: A Case in the SADC Region

In 2002, in response to the worst food crisis ever experienced in southern Africa that put 14 million people in a state of extreme starvation, food aid shipments from the USA containing genetically modified (GM) crops were dispatched to address the situation. However, the governments of Malawi, Mozambique, Zambia and Zimbabwe rejected this aid due to concerns over the inclusion of GM maize in the consignment. This led to much debate by corporates, governments and biotechnology critics, in and outside the region, over GM food aid, specifically in relation to concerns about the health and environmental dangers they posed.

The Biotechnology Outreach and Information Transfer Project (2002–2005) was launched in response to the apparent void of information on biotechnologies in the SADC region—considered to be responsible for the rejection of GM food by the region during the 2002 food crisis. This was an initiative of USAID and implemented by AfricaBio, a South African biotechnology stakeholder association. The objective of the project was to conduct comprehensive biotechnology advocacy. The project succeeded in linking public and private sector institutions, researchers and policymakers in the region to learn about biotechnology and its potential to rid the continent of hunger. Project impacts included:

- Establishment of demonstration plots in South Africa to showcase and compare production of Bt[1] and non-Bt maize to smallholder farmers in the region. Some were convinced by the results to consider planting the Bt variety.
- The Biotechnology Outreach Society of Zambia (BOSZ) trained journalists in effective science communication, which resulted in a rise in the number of positive stories on biotechnology. Yatsani Radio, a community station, interviewed members of the public on their understanding of biotechnology. They found that the public yearned for more information on the subject.
- BOSZ representatives participated in consultative meetings on biosafety and subsequently held talks with members of parliament to push for the approval of the draft biotechnology policy. Advocacy efforts from BOSZ and other similar groups led to the enactment of the Biosafety Act of Zambia in 2007.

(continued)

---

[1] Bt maize refers to genetically engineered maize varieties that produce crystal (Cry) proteins or toxins derived from the soil bacterium, *Bacillus thuringiensis*.

Box 3.1 (continued)

- The project hosted a delegation of government officials from Angola for a fact-finding mission to get information on how to develop and implement a national biosafety framework. Following this, Angola developed its Biosafety Bill in 2005.
- A comprehensive database of decision makers, regulators and scientists was developed. In addition, exhaustive profiles of anti-GM campaigners were developed. The latter were included in project events intended to display to them the benefits GMOs.

The overall impact of this initiative has been an increased public awareness and willingness of some governments in the region to consider agricultural biotechnology as an important means of addressing food production and availability challenges.

Source: Compiled by authors

At the international level, global efforts with African engagements include the Program Biosafety System, which supports African countries to develop, implement and manage their own systems by providing training, technical and legal advice and independent policy research for decision makers. It has also developed tools for strategic and systematic outreach to create awareness among stakeholders.

The United Nations Environment Programme-Global Environment Facility (UNEP-GEF) initiative on supporting countries to develop and implement national biosafety frameworks, as well as facilitating subregional and regional information sharing and dialogue, has contributed significantly to creating enabling environments at political levels (Box 3.2).

Box 3.2 A Bilateral Approach to Capacity Building for Creating Biotechnology Enabling Environments: The UNEP-GEF NBF Project

Several SSA countries have received assistance under the UNEP-GEF project on the development of national biosafety frameworks (NBFs). This global project began in June 2001. It was designed to assist countries to develop their NBFs in order to comply with the Cartagena Protocol on Biosafety.

The NBF project developed a set of core activities and created a toolkit (available in four languages) to help countries develop draft NBFs, which generally included plans for a government policy on biosafety, a regulatory framework, administrative structures to handle requests or applications for decisions on living modified organism (LMO) handling and transfer, a system for increasing public awareness and promoting public participation in decision making and enforcement and monitoring systems. An additional component

(continued)

Box 3.2 (continued)

involved promoting collaboration and exchange of experiences on biosafety among countries.

As part of the national-level component, national surveys were carried out to identify existing applications of modern biotechnology; the extent and impact of releases of LMOs, biosafety, risk assessment and risk-management systems; and reviews of existing legislation relevant to biosafety.

By 2012, 37 SSA countries had been assisted to develop their NBFs while 14 had been assisted to implement the frameworks. The project convened five regional and subregional workshops, and smaller meetings focused on subregional collaboration. These meetings and workshops helped national project staff to increase their knowledge and to catalyse cross-country learning. They also promoted south-south collaboration and networking, with countries increasingly requesting technical assistance from other developing countries that have done similar work. Outcomes in each participating country have included:

- A policy on biosafety
- An operational regulatory regime
- Workable and transparent administrative system
- Workable and transparent systems for public information, public participation and enforcement and monitoring
- Enhanced technical and laboratory capacity for LMO detection
- National websites and/or national Biosafety Clearing-Houses

An evaluation at the closure of the project in 2016 concluded that there were mixed results and that the initial asymmetry between countries was an important factor. Some 25–30% of the countries moved steadily towards NBF implementation and to higher levels of results (improved decision making and biosafety governance at national level), and another 25–30% stayed well behind (with no significant steps towards NBF implementation), whereas the majority of the countries (40–50%) somewhat progressed in setting the NBF (e.g. a national law, National Competent Authority in place), yet these countries cannot claim to have it fully operational due to evident flaws (e.g. lack of regulations and administrative procedures, insufficient institutional uptake and stakeholder participation).

Source: CBD (2013), UNEP (2006, 2017)

At the regional level, NEPAD, through its programme of the African Biosafety Network of Experts, has been providing training and public awareness in policy, legislation and regulation on biotechnology with a focus on GMOs for countries in Africa (Makinde et al. 2009). Other more independent frameworks, such as AfricaBio, African Biotechnology Stakeholders Forum (ABSF) and the Open

Forum on Agricultural Biotechnology (OFAB), provide platforms for continuous dialogue by biotechnology stakeholders, strengthening interinstitutional networking and sharing of biotechnology information and promoting interactions between scientists, journalists, the civil society, industrialists and policymakers. Some examples of subregional efforts include Réseau des Communicateurs ouest-Africains sur la Biotechnologie (RECOAB), which is a network of both francophone and anglophone West African journalists that builds capacity and provides factual and balanced information on biotechnology to enable informed participation in debates on biotechnology (Chambers et al. 2014; Olembo et al. 2010).

At the national level, biotechnology communication, public awareness and participation are undertaken by both governmental and non-governmental initiatives. For example, in Kenya the National Biotechnology Awareness Creation Strategy (BioAWARE), an initiative by the National Council for Science and Technology (a semi-autonomous government agency—now the National Commission for Science Technology and Innovation), has contributed significantly to raising awareness and providing platforms for public engagement on biotechnology matters. Similar initiatives exist in other countries including Public Understanding of Biotechnology (PUB) in South Africa and the Burkina Biotech Association. Table 3.2 summarizes the key biotechnology networks and outreach programmes in SSA.

3.2.2   Biotechnology Policy and Biosafety Frameworks

There are several intergovernmental mechanisms that govern the application of modern biotechnology in which African countries actively participate. Key among these are the International Plant Protection Convention, the Codex Alimentarius Commission and the Cartagena Protocol on Biosafety to the Convention on Biological Diversity (Makinde et al. 2009; Chambers et al. 2014; Gordh and McKirdy 2014). At the national level, national biosafety frameworks (NBFs) provide the overall policy, legal and institutional mechanisms for development, deployment and use of biotechnology. The frameworks comprise the biosafety laws and policies, guidelines and regulations, as well as the regulatory institutions (UNEP 2000, 2006; Karembu et al. 2010).

Enabling movement of germplasm, especially seed, between countries is a critical factor in driving varietal development and availability of competitively priced quality seed to farmers. Despite recognition of need for harmonization among countries and ongoing efforts within the regional economic communities (RECs), this has—by and large—still not been achieved. There is urgent need for a coordinated regional approach so that a variety tested in one country could be added to the list in another or could be registered after a short period of testing (e.g. 1 year). Where harmonization processes have been operationalized, the harmonized lists of varieties in the varietal registers of most countries should be regularly updated and efforts made to assist countries that are lagging behind. For this to happen in the majority of RECs, new rules and standards must be incorporated into national law in order to make regional initiatives effective. Understanding national rules that govern

Table 3.2 Key biotechnology networks, initiatives and outreach programmes in SSA

Programme or initiative	Area of focus
International Service for the Acquisition of Agri-biotech Applications (ISAAA) AfriCenter	AfriCenter focuses on communication and knowledge sharing through collection, packaging and dissemination of knowledge and through networking, building partnerships and fostering joint initiatives to share resources, experiences and expertise on crop biotechnology. It coordinates a network of biotechnology information centres (BICs) located in Egypt (EBIC for Arab-speaking people), Mali and Burkina Faso (Mali-BIC for French-speaking) and East and Central Africa (ECABIC for English- and Swahili-speaking). www.isaaa.org/kc
Public Understanding of Biotechnology (PUB)	PUB is operated by the South African Agency of Science and Technology Advancement (SAATA) with the aim of promoting a clear understanding of biotechnology's potential and ensuring broad public awareness to stimulate dialogue and debate on biotechnology. www.pub.ac.za
AfricaBio	AfricaBio is a biotechnology stakeholders' association whose key role is to provide accurate information and create awareness, understanding and knowledge on biotechnology and biosafety in Africa. www.africabio.com
African Biotechnology Stakeholders Forum (ABSF)	ABSF's focus is on creating an innovative and enabling biotechnology environment in Africa through education, enhanced understanding and awareness creation. www.absfafrica.org
Africa Harvest Biotech Foundation International (Africa Harvest)	Africa Harvest aims to build the capacity of scientists and of science and agricultural organizations in integrating communication strategies into their research activities and also to help the news media improve their coverage of science and agricultural issues. www.ahbfi.org
National Biotechnology Awareness Creation Strategy (BioAWARE-Kenya)	Spearheaded by the Agricultural Sector Coordinating Unit (ASCU) under the Ministry of Agriculture, BioAWARE-Kenya aims to provide a knowledge base for informed decision making to hasten the deployment of biotechnology through a participatory awareness creation process
Open Forum on Agricultural Biotechnology (OFAB) in Africa	OFAB's focus is on strengthening interinstitutional networking and sharing of credible, sound and factual biotechnology information through a platform that brings together stakeholders in biotechnology and enables interactions between scientists,

(continued)

Table 3.2 (continued)

Programme or initiative	Area of focus
	journalists, the civil society, industrialists and policymakers. www.ofabafrica.org
Réseau des Communicateurs ouest-Africains sur la Biotechnologie (RECOAB)	RECOAB is a network of both francophone and anglophone West African journalists that builds capacity and provides factual and balanced information on biotechnology to enable informed participation in debates on biotechnology
Biotechnology-Ecology Research and Outreach Consortium (BioEROC)	BioEROC, based in Malawi, aims to deliver relevant research, training, management and outreach services in natural resources to promote responsible and relevant applications of biotechnology and its products
Burkina Biotech Association	Burkina Biotech Association was created by Burkina Faso scientists with the objective of providing a forum for stakeholders in the field of biotechnology to dialogue and voice their opinions and concerns

Source: Karembu et al. (2009)

variety release and registration, seed certification, trade and sanitary and phytosanitary measures, as well as plant variety protection laws within the context of regional harmonization, will assist in understanding how implementation of regional rules will work in practice (Kuhlmann 2015).

It is also important that differences between regions, including differences in regional legal frameworks, are well understood. For instance, the Common Market for Eastern and Southern Africa (COMESA), which approved seed regulations in 2014, is structured such that regional rules are binding, but national-level implementation is required before such rules can take effect. Under SADC, only protocols are legally binding, while other SADC instruments, including memorandums of understanding, are not. This state of affairs is problematic for countries which are members of the two RECs. An examination of the status of COMESA countries as part of the harmonization process, for example, classifies member states into three categories that signify readiness to implement the 2014 COMESA seed regulations: (1) countries with existing legal structures: Egypt, Eswatini, Ethiopia, Kenya, Madagascar, Malawi, Sudan, Uganda, Zambia and Zimbabwe; (2) countries with legal structures in draft form: Burundi, DRC, Mauritius, Rwanda and Seychelles; and (3) countries with no legal structures: Comoros, Djibouti, Eritrea, Libya and South Sudan. An ongoing initiative covering six countries in COMESA and the East African Community (EAC) regions (Kenya, Uganda, Tanzania, Ethiopia, Rwanda and Zambia) and funded by the Alliance for a Green Revolution in Africa (AGRA) is working with seed stakeholders in these countries (including the national regulatory authorities) to put in place two tools as a way of enabling seed movement across borders of countries in the region.[2] One of the tools is an online information guide being developed to facilitate ease of access to critical seed trade requirements so that seed traders have the same information that authorities (including customs officials) have and can confidently engage these authorities and reduce corrupt practices based on fabricated requirements. A second tool being introduced is an audit mechanism through which national seed certification processes are voluntarily audited by seed inspectors from other countries in the region to create an environment of mutual trust and accountability that will increase acceptability of seed across borders from neighbouring countries.

Sub-Saharan African RECs are at different stages in the establishment of variety catalogues, an important instrument for operationalizing harmonization. Any variety entered into the national catalogue of a member state should be mirrored on the regional catalogue and be freely traded and allowed for multiplication throughout the region without any further registration requirement. The Economic Community of West African States (ECOWAS) differs from other regions in that new varieties only need to be registered in one member country in order to be eligible for entry in the regional catalogue. COMESA and SADC require registration in two countries in order to be eligible for entry in the regional catalogue.

Biosafety Frameworks A national biosafety framework (NBF) is a combination of policy, legal, administrative and technical instrument that is set in place to address safety for the environment and human health in the context of modern biotechnology (Wafula et al. 2012). These frameworks often focus on GMOs and have been generally driven by the crop sector, although they are meant to cover broader biotechnology research and applications.

Although NBFs vary from country to country, they usually contain a number of common elements (Karembu et al. 2010). These include policy on biosafety, often as part of a broader policy on biotechnology; regulatory regime for biosafety including an act and regulations; and a system to handle notifications or requests for authorizations for certain activities, such as registration of activities (contained use) and field releases of GMOs into the environment. This involves public participation and risk assessment, a mechanism for monitoring and inspections and a system for public awareness and public information, i.e. how to inform stakeholders about the development and implementation of the NBF (UNEP 2000, 2006).

Today, biosafety regulatory systems constitute an important determinant for the introduction and deployment of biotechnology, especially GMOs. This gained momentum with the coming into existence in 2003 of the Cartagena Protocol on Biosafety at the international level that in part guides the establishment of national biosafety regulatory systems. As at 2020, nearly all SSA countries had ratified or complied with accession requirements of the Biosafety Protocol (CBD 2000) with the exception of Equatorial Guinea and South Sudan. The establishment of national biosafety frameworks builds on substantial support provided under the UNEP-GEF

---

[2]https://www.comesa.int/working-with-agra-to-fast-track-seed-harmonization/

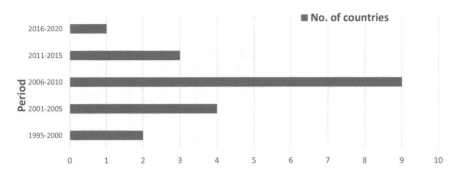

Fig. 3.1 Trends and status of biosafety legislation in SSA

and complemented by other international and regional actors in capacity building efforts. However, the road to the establishment of NBFs across the continent has been long and winding (UNEP 2000). While a few countries stand out as champions, a large majority are in different stages of putting in place the various components of functional NBFs. At the same time, countries are benefiting from regional integration activities being advocated for by the African Union and main economic blocs to seek common mechanisms that would complement NBFs. For example, most of the countries in the eastern and southern Africa subregion belong to COMESA. They developed their biosafety frameworks under the UNEP-GEF project during 2001–2004 (UNEP 2006; Karembu et al. 2010; Chambers et al. 2014: Makinde et al. 2009).

The levels and extents of development of the frameworks among SSA countries largely depend on their adherence to and domestication of key international agreements, political goodwill as well as both human and financial capacities. As summarized in Fig. 3.1, SSA countries started putting in place biosafety legislation in the 1990s; today, only 19 countries have biosafety legislation in place. The majority of these (nine) were passed in the period 2006 to 2010. The country-specific status on the development and implementation of NBFs, regulations and institutions is provided in Table 3.3.

3.2.3 Public and Private Sector Investments

The basic justification for government expenditure on agricultural research is that in the absence of public intervention, private firms will underinvest in research when the output of that research has the characteristics of a public good. The case was made by Ruttan (2000) who wrote:

New information or knowledge resulting from research is typically endowed with the attributes of a public good characterized by non-rivalness or jointness in supply and use and non-excludability or external economies. The first attribute implies that the good is equally available to all. The second implies that it is impossible for private producers to appropriate through market pricing the full social benefits arising directly from the

Table 3.3 Current status of national biosafety frameworks in SSA countries

Country	Biosafety act	Biosafety legislation, regulation policy	Regulatory agency
Botswana	Bill under review	Nat Biosafety Policy 2013	
Burkina Faso	Act 2006; revised 2013	Biosafety Decree 2004; Biosafety Law 2011; Policy on Biotech	National Biosafety Authority
Cameroon	Biosafety Act 2003; revised 2007	Biosafety guidelines 1995; draft	
Eswatini	Biosafety Act 2012	Legislation under review	
Ghana	Act 2011; enacted law 2012	Regulatory framework yet to be finalized; Policy on Biotech; Regulatory Communication Strategy 2014	National Biosafety Committee
Kenya	Act 2009	Regulation and Guidelines 2011; National Biotechnology Policy 2006	National Biosafety Authority
Lesotho	National Biosafety Bill 2005; amended 2014	National Biosafety Policy; National Biosafety Awareness Strategy 2013; draft legislation	National Biosafety Council under the Ministry of Tourism, Environment and Culture
Malawi	Act 2002	Biosafety guidelines 1995; Biosafety Regulatory Framework 2007; National Biotech Policy 2008; draft legislation	Department of Environmental Affairs, Ministry of Environment and Climate Change Management
Mali	Biosafety law 2008; review of biosafety decree 2010		National Biosafety Committee
Mauritius	GMO Act 2004 Plant Protection Bill 2006		Ministry of Agro Industry and Food Security
Mozambique	Biosafety law 2007 revised 2012 bill under review	Draft biosafety regulations/guidelines	Grupo Inter-Institucional Sobre BioSegurança (GIIBS); NBC
Namibia	Biosafety Act promulgated 2006	Draft legislation Biotechnology and biosafety policy 1999	Biosafety Council of the National Commission on Research, Science and Technology (NCRST)
Nigeria	Biosafety bill 2011. Bill passed second reading Jul 2014,	Biosafety guidelines 2001	Federal Ministry of Environment; National

(continued)

Table 3.3 (continued)

Country	Biosafety act	Biosafety legislation, regulation policy	Regulatory agency
	referred to Senate Committees on Agriculture, Science and Technology		Biosafety Management Agency
Senegal	Biosafety law 2009		Ministry of Environment
South Africa	GMO Act 1997	Biosafety guidelines; National Biotechnology Policy and Strategy 2001	Directorate of Biosafety
Sudan	Law of Biosafety 2010	National Biosafety Framework 2008	Sudan National Biosafety Council
Tanzania	Environment Management Act of 2004	Biosafety regulation 2009; biotech policy 2010; policy under review	National Biosafety Committee
Togo	National Biosafety Framework 2004; Biosafety Law 2009	Draft legislation	Ministry for Environment and Forestry, National Biosafety Committee
Uganda	National Biosafety Bill 2012; passed 2017, not yet assented into law	Biosafety guidelines 1995; Draft Biotech and Biosafety Policy 2013; National Biotechnology Policy 2008	Uganda National Council for Science and Technology; National Biosafety Committee
Zambia	Biosafety Act 2007, reviewed 2013	Nat Biosafety Policy 2013; Nat Biosafety Body 2013	National Biosafety Authority
Zimbabwe	National Biotech Authority Act 2000	Biosafety guidelines 1998	National Biotechnology Authority of Zimbabwe

production (and consumption) of the good—it is difficult to exclude from the use of the good those who do not pay for it. A socially optimal level of supply of such a good cannot be expected if its supply is left to private firms.

In spite of the economic relevance of agriculture and its substantial contribution to the economies of most African countries, the proportion of public expenditures that governments devote to agriculture (not just research) does not reflect this. Data for 20 countries in Africa compiled by IFPRI in 2016 covering 2005–2011 (more recent for some countries) showed that only 5 countries nearly reached or surpassed the Maputo Declaration target of allocating 10% national budgets to agriculture. These were Burkina Faso, Ghana, Malawi, Rwanda and Senegal. In addition, during this period, Sierra Leone more than doubled its share from 3 to 7.5%. For eight countries, the proportion of total budgets spent on agriculture declined, and for four countries, the actual expenditure also declined (Mink 2016).

Biotechnology R&D is a resource-intensive endeavour. With respect to investment specifically in agricultural biotechnology, an accurate analysis of the financial picture for Africa is difficult to achieve due to the lack of information. Some data from the FAO Global Partnership Initiative for Plant Breeding Capacity Building (FAO 2021) on trends in public-sector expenditures for plant breeding and biotechnology shows that plant breeding budgets for all African countries with available data decreased dramatically, from USD347 million in 1985 to USD99 million in 2005. Although not indicative of biotechnology expenditures, these budget trends reveal diminishing financial support to the breeding platform that is needed to support biotech interventions.

International development partners through the CGIAR centres, NGOs and other initiatives bridge some of the investment gaps for many countries related to supply of genetically pure stock materials of preferred local variety for bulking and further multiplication; facilitation of diffusion of varieties developed through participatory varietal selection; training of community seed producers and farmers on quality issues related to production; and assistance with marketing or distribution of seed within and between communities. The absence or limited presence of private sector players in the seed sector in many countries—especially West and Central Africa— is also a major problem, related in significant ways to the regulatory environment. Small-scale seed enterprises are beginning to emerge and are enhancing access to quality seeds by farmers. Having access to a regional seed market with fewer barriers to seed trade would further contribute to their development.

There are no recent statistics on public spending in agricultural biotechnology for most countries. However, public spending in agricultural R&D since the Maputo Declaration offers a more optimistic picture and a step in the right direction. In 2011, SSA countries spent USD1.7 billion on agriculture R&D, an increase of more than 30% over the USD1.2 billion recorded in 2000 (Beintema and Stads 2014).

Actual spending on agricultural biotechnology will be a small proportion of the overall spending on agriculture. Recent reliable data is generally lacking, but Mugabe (2002) reported that this amounted to less than USD250,000 per year for most countries. In Nigeria, however, the National Agriculture Biotechnology Development Agency was reported in 2014 to be providing USD263 million per year in start-up funding for biotechnology (Chambers et al. 2014).

In a few SSA countries, for example, Kenya and South Africa, both domestic and multinational companies are getting involved in private sector R&D. However, even where private sector investment in agriculture biotech research is occurring, adoption by small-scale farmers is usually limited by weak or absent extension services and lack of ready access to the necessary inputs. In addition, often the available technology is not well adapted to local conditions (Pray et al. 2007). Willingness to pay is also a factor as seeds of GM crops developed by the private sector are more expensive than conventional seeds and need to be bought each season. Manoeuvring the intellectual property landscape is also cumbersome for private companies, acting as a significant disincentive to market entry.

Public-private partnerships (PPPs) have been shown to present a major opportunity for harnessing synergies and ensuring efficiency, effectiveness and sustainability of innovations and are emerging in many African countries and are already quite visible in a few. PPP as a model can be effective in ensuring that private research, for example, on GM crops, is integrated with knowledge of varieties and local cropping conditions from the national agricultural research systems (NARS). This can result in the development of GM crops that are appropriate to African conditions and needs. Agricultural R&D undertaken through PPPs can include genetic improvements of crops, complementary approaches to seed systems, agriculture extension, financial and infrastructure development as well as market access (GCARD 2010). Examples of PPPs in the crop biotech domain in SSA include R&D work on beans done under the Pan-African Bean Research Alliance (PABRA), the development and scaling of Aflasafe™ and the biofortified banana initiative in Uganda. Details are provided in Sect. 4.5 of Chap. 4 (Crop Impact Cases).

However, in view of the fact that such PPPs have been primarily fostered by donors, it is crucial that design is deliberate in ensuring sustainability. Past lessons (e.g. Muraguri 2010) suggest that such donor-funded partnerships do not survive beyond the end of funding. Ultimately, productive and sustainable partnerships will be those that emerge from felt needs by, and mutual benefits for, the public and private sectors without donor-type financing.

Figure 3.2 presents the status of agricultural R&D spending by SSA countries based on ASTI data as of 2014. South Africa, Nigeria, Kenya, Ghana and Uganda were the top spenders in terms of total agricultural R&D and biotech (biotech figures for South Africa were not available). As was the case with agricultural R&D staffing, spending figures show that Mauritius and Botswana are high investors on a per million inhabitants basis. South Africa, Eswatini and Cape Verde also rank high on this metric. While these figures covered agricultural R&D broadly, a major component is for crops, so that the ranking of spending by countries on crops correlates closely with total agricultural R&D spending.

3.2.4   Classification of Countries

On the basis of the above analysis, SSA countries were grouped into five categories based on overall status of enabling environment for crop biotechnologies (Table 3.4).

3.3   Livestock

3.3.1   Public Awareness and Political Support

Several countries have specific initiatives and policies to promote livestock production through application of biotechnologies, particularly for control of diseases, improvement of feeds and introduction and improvement of livestock genetics. Many countries, however, still do not have specific livestock policy documents, and livestock is often covered under agricultural policy. A few countries do have

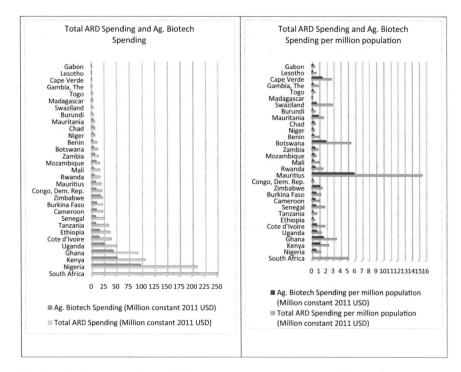

Fig. 3.2 Total agricultural R&D (ARD) and biotech spending (USD) in SSA—absolute and per million inhabitants (2014)

policies specific to livestock. Examples of the latter are Ethiopia's Livestock Development Master Plan 2015–2020 (Shapiro et al. 2015) and the Tanzania government's Livestock Policy 2006 and Livestock Modernization Initiative (2015) (MLD 2006; MLFD 2015).

These policies, plans and initiatives have the objective to improve the productivity and health of livestock, including the identification of specific biotechnology applications in a range of species and with specific objectives (nutrition, animal health, genetic improvement) so as to meet the rising national and regional demand for livestock products. Countries differ in terms of both the extent to which the role of livestock technology is explicitly identified and the purpose.

Ethiopia's Livestock Master Plan, for example, makes no explicit reference to application of modern livestock biotechnology although our review of biotechnology use in the country identifies significant research and applications in livestock. Kenya's National Livestock Policy (Session Paper No. 2 of 2008) and more recent (2017) conversations towards a revised policy indicate a desire for increased application of biotechnologies in genetic characterization, conservation and improvement as well as in improving nutrition and animal health. Cameroon is undertaking a large programme to revitalize its livestock sector for improved productivity and resilience

Table 3.4 Classification of countries on the basis of enabling environment for application of biotechnology in crops

Enabling environment category	Countries
Very weak	Angola, Burundi, Chad, Central Africa Republic, Congo, Djibouti, DRC, Eswatini, Gambia, Equatorial Guinea, Gabon, Guinea, Guinea Bissau, Lesotho, Liberia, Niger, Sierra Leone, Togo, Somalia, South Sudan
Weak	Benin, Burkina Faso, Cameroon, Côte d'Ivoire, Eritrea, Madagascar, Mozambique, Rwanda, Senegal
Medium	Botswana, Ghana, Malawi, Mali, Namibia, Tanzania, Uganda, Zambia, Zimbabwe
Strong	Kenya, Nigeria, Sudan, Ethiopia
Very strong	South Africa

Countries not listed due to lack of sufficient data: Cape Verde, Comoros, Mauritania, Mauritius, Sao Tome and Principe and Seychelles

to climate change (World Bank 2016), while Mali is building on the major artificial insemination initiative, which was established in 2014.

3.3.2 Biotechnology Policies and Biosafety Frameworks

As is the case in crops, the livestock industry in many SSA countries is regulated by specific national laws and regulations. In livestock, these are administered by the livestock production and veterinary departments. The sector is also subject to other laws and regulations that relate to food safety, public health and the environment. Generally, livestock biotechnology is controlled through the animal diseases acts (e.g. Kenya, Tanzania, Ethiopia) or equivalent laws, the animal breeding acts (e.g. Uganda and South Africa) and, more recently, the biotechnology and biosafety laws (e.g. Burkina Faso and Nigeria 2006). Furthermore, international conventions and protocols (e.g. the CBD, the Cartagena Protocol and Nagoya Protocol) and agreements and guidelines (e.g. the African Model Law on Safety in Biotechnology 2007) provide and inform the legal and regulatory frameworks for development and use of livestock biotechnologies.

Currently all SSA countries have laws and regulations which to varying degrees touch on the conventional, mostly crop-focused biotechnologies. Many of these laws are a legacy of the immediate postcolonial period and, in many countries, are overdue for revision to take into consideration advances in biotechnology. Most SSA countries have ratified the CBD and associated protocols, but not all have enacted biotechnology and biosafety laws (see section on crops, this Sect. 3.2.2). Moreover, in many of the countries in which the biosafety laws have been enacted, the process was driven by a crop-centric narrative aimed specifically at the regulation of genetically modified crops and have had to be retrofitted with regulations to govern animal biotechnology. Also, with the exception of South Africa and Burkina Faso, the biosafety laws and frameworks adopted by many SSA countries are modelled on the African Model Law 2007 and underpinned by the precautionary principle. Consequently, they are rather restrictive to development or adoption of modern biotechnology in animals. In contrast, the South African biosafety framework predates the African Model Law, while the Burkina Faso biosafety law was responsive to prior introduction of *Bt* cotton and takes a product-based approach to regulation (Godfrey 2013).

Development of an effective regulatory system for genetically engineered animals and their products has been the subject of increasing discussion among researchers, industry and policymakers, as well as the public. Since transgenesis and cloning are relatively new scientific techniques, transgenic animals are new organisms for which there is limited information. The issues associated with the regulation and biosafety of transgenic animals pertain to environmental impact, human food safety, animal health and welfare, trade and ethics (Kochhar et al. 2005). To regulate this new and powerful technology, predicated on limited background information, is a challenge not only for the regulators but also for the developers of such animals, who strive to prove that the animals are safe and merit bioequivalency to their conventional counterparts. However, like other regions of the world, questions around biotech applications in animals revolve around possible risks, such as expression of novel proteins in meat, milk or eggs that could cause allergic reactions in susceptible people and the potential escape of transgenic animals, although the absence of wild relatives of most livestock species in most regions minimizes the latter threat.

As in crops, where a major challenge of immediate concern to a majority of farmers is access to quality affordable inputs, especially seeds, the same applies to livestock. In addition to feeds, access to animal health services is a major bottleneck to the use of superior genetics. Although most SSA livestock are still predominantly dependent on pastures, it is important to note that genetically engineered crops will increasingly benefit the livestock sector through increased yields of feed ingredients and enhanced feed quality traits. The relevant crops used in livestock feeds include feed maize, canola (rapeseed), cottonseed and soybean. These crops are principally used in livestock feed rations in intensive livestock systems either as an energy or protein source.

Most countries have a system of assessment and approval of veterinary medicines to ensure that they meet high standards of safety, quality and efficacy before they are authorized for sale (i.e. registration). However, these systems are not functional in many countries; in some countries there are no such systems in place for registering medicines, while in countries that do have a registration system, it is often designed for human medicines, and there are no distinctions made between requirements for pharmaceuticals and vaccines (GALVmed 2015). Assessing applications under these circumstances, without clear, specific guidelines, can lead to long delays for applicants seeking marketing authority and presents a challenge especially for movement of veterinary vaccines.

The Global Alliance for Livestock Veterinary Medicines (GALVmed), an international not-for-profit public-private partnership, has been working with the East African community to put in place a mutual recognition procedure to avoid duplication of assessments of the same product by different regulatory authorities and therefore facilitate movement, for example, of vaccines across borders.

3.3.3 Public and Private Sector Investments

Like any sector, the development of the livestock biotechnology sector requires investments in R&D and dissemination of the technologies to the end users. In all SSA countries, the biggest source of funding is through public sources (government, principally through loans and grants from development partners) with a small fraction coming through the private sector.

Data on budget expenditures for livestock R&D by countries was not available. However, the ASTI initiative of IFPRI has collected expenditure data on agriculture R&D (IFPRI 2014). With a few exceptions, like Namibia which spent 3% of its agricultural GDP in 2014 on R&D, most SSA countries spent less than 2% of the value of agricultural contribution to their GDP in the same year. When this expenditure is expressed as a ratio of the human population, Kenya and South Africa have high R&D expenditure—comparable to those of developed countries (Nin-Pratt 2016). It is not clear from the ASTI data the relative content of the expenditure on livestock (versus crops and other sectors of agricultural) R&D and, further, what proportion of that goes towards biotechnology. However, assuming that resource allocation correlates with full-time (staff) equivalents (FTEs) then by using the proportion of FTE allocated to livestock research, we can get insights into the proportion of resources allocated to overall livestock R&D. Using this approach, investment in livestock R&D in SSA is generally between 10 and 15% of the overall investment in agricultural R&D. Furthermore, the share of investment in livestock R&D is far lower than the contribution of the sector to agricultural GDP. For example, although livestock contributes more than 50% of Namibia's agricultural GDP, the sector has only 25% share of the R&D resources.

The AU-IBAR programmes relevant for biotechnology, e.g. Pan-African Veterinary Vaccine Centre (AU-PANVAC) in Ethiopia, and the ongoing pan-African project on animal genetic resources characterization, Strengthening the Capacity of African Countries to Conservation and Sustainable Utilization of African Animal Genetic Resources, provide additional investments across the continent and serve as important catalysts in countries which host some of the regional facilities. The same can be said about countries hosting OIE reference labs, e.g. Botswana which hosts the contagious bovine pleuropneumonia reference laboratory.

Complementary to government public spending is investment by non-governmental organizations which provide resources to promote uptake of biotechnologies by livestock farmers. For example, the Bill and Melinda Gates Foundation (BMGF) is investing tens of millions of dollars in the development and dissemination of biotechnologies for poultry, genetic improvement of cattle and animal health. Some of the BMGF-funded initiatives are the Centre for Tropical Livestock Genetics and Health in collaboration with ILRI that aims to catalyse development of high-level biotechnologies for poultry and dairy cattle for the tropics; the Dairy Genetics East Africa project led by ILRI (between 2012 and 2005) which aimed to identify dairy genotypes appropriate for smallholders; the African Chicken Genetics Gain Project also implemented by ILRI (and concluded in 2019) in Nigeria, Ethiopia and Tanzania which tested a range of global chicken genotypes for appropriateness to smallholder settings; the Public-Private Partnership for AI Delivery (PAID) in Ethiopia and Tanzania (Land O'Lakes International Development 2015) to promote the participation of the private sector in sustainable delivery of superior genetics for dairy cattle using AI (scheduled to end in 2021); and GALVMed, enhancing the capacity (including regulatory capacity) to develop, manufacture and deliver critical veterinary vaccines in Africa.

Worthy of note here is ILRI and the ILRI-Biosciences eastern and Central Africa (ILRI-BecA) Hub. The ILRI-BecA Hub remains the single major state-of-the art platform for research on livestock biotechnology in SSA—with a focus on genetic characterization, genetic improvement and animal health—especially diagnostics and vaccine research.

Inadequate investment in rural infrastructure, especially all-weather roads and power supply, represents one of the major challenges to last-mile delivery of technologies—for example, AI and animal health services, especially vaccines which require cold chain—in all SSA countries.

3.3.4 Classification of Countries

The state of enabling environment in SSA countries was assessed using the following three broad criteria: existence of a biosafety framework—incorporating policy and the extent to which this recognizes the potential of biotechnology and promotes improvement in the biological efficiency of livestock through such technologies; acts of parliament, legislation, regulation and existence of a specific agency with responsibility for implementation of the framework; extent to which the framework has been operationalized; and level of investment in livestock biotechnologies.

While quantitative data was not available, information from survey and desk reviews was used to assess level of livestock biotechnology R&D in individual countries and processes in place for programme approvals and execution as indicators of the operationalization of the regulatory environment in each country. Similarly, while the assessment did not gather explicit data on level of public investment, we used, in addition to actual budget figures, where available, numbers and scopes of R&D projects and programmes as proxy measures of the relative levels of investment.

Table 3.5 presents a summary of country clusters into the five categories of enabling environment. The table shows the dire situation in SSA in relation to enabling environment for biotech applications in livestock—almost identical to the picture seen in crop sector. The majority of countries are in the 'very weak' (21) and 'weak' (8) categories, with only ten in the 'medium' category. Although improvements (especially in relation to investments) are still needed in all countries, Kenya and to some extent Nigeria ('strong') and South Africa ('very strong')

Table 3.5 Classification of countries on basis of enabling environment for applications of bio-technology in livestock

Enabling environment	Countries
Very weak	Angola, Benin, Burundi, Chad, Central African Republic, Republic of Congo, Djibouti, DRC, Gambia, Equatorial Guinea, Eritrea, Eswatini, Gabon, Guinea, Guinea Bissau, Lesotho, Liberia, Niger, Sierra Leone, Togo, Somalia, South Sudan
Weak	Burkina Faso, Cameroon, Côte d'Ivoire, Madagascar, Malawi, Mozambique, Rwanda, Senegal
Medium	Botswana, Ethiopia, Ghana, Mali, Namibia, Sudan, Tanzania, Uganda, Zambia, Zimbabwe
Strong	Kenya, Nigeria
Very strong	South Africa

Countries not listed due to lack of sufficient data: Cape Verde, Comoros, Mauritania, Mauritius, Sao Tome and Principe and Seychelles

currently have, overall, better enabling environment for livestock biotechnology applications. Responses obtained from national authorities differed from the classification done based on the framework applying all the information from desk reviews and other sources. For example, officials from Botswana, Ghana and Zimbabwe suggested in their response that the enabling environment in these countries is 'weak', while Nigeria suggested that theirs falls in the 'very weak' category. As summarized in Table 3.3, these suggestions are at variance with what available information seems to reveal.

References

Beintema N, Stads GJ (2014) Is Africa investing enough? In IFPRI 2013 global food policy report. International Food Policy Research Institute, Washington, DC

CBD (2000) Cartagena protocol on biosafety to the convention on biological diversity: text and annexes. Secretariat of the Convention on Biological Diversity, Montreal. Available via CBD: https://www.cbd.int/doc/legal/cartagena-protocol-en.pdf Accessed 30 June 2021

CBD (2013) Biosafety clearing house. Secretariat of the Convention on Biological Diversity, Montreal. Available via CBD: https://bch.cbd.int/protocol/gefprojects.shtml. Accessed 30 June 2021

Chambers JA et al (2014) GM agricultural technologies for Africa: a state of affairs. Report of a study commissioned by the African Development Bank. IFPRI and AfDB, Washington and Abidjan

FAO (2021) Global partnership initiative for plant breeding capacity building. Available via FAO. www.fao.org/in-action/plant-breeding/background/zh/. Accessed 30 June 2021

GALVmed (2015) Review of requirements and processes for veterinary products in selected African and Asian countries. GALVmed, Edinburgh. Available at https://assets.publishing. service.gov.uk/media/5aa67844e5274a3e391e38bf/58_Vaccine_registration_report_final_ draft.pdf. Accessed 7 October 2021

GCARD (Global Conference on Agricultural Research for Development) (2010) Benefitting the poor through public-private partnerships for innovation and action. Background paper for the global conference on agricultural research for development. Montpellier, March 28–30

Godfrey RN (2013) Case studies of African agricultural biotechnology regulation: Precautionary and harmonized policy-making in the wake of the cartagena protocol and the AU model law. Loy LA Int'l & Comp L Rev, 35, p 409. Available at https://digitalcommons.lmu.edu/ilr/vol35/iss3/3

Gordh G, McKirdy S (2014) The handbook of plant biosecurity: principles and practices for the identification, containment and control of organisms that threaten agriculture and the environment globally. Springer. https://doi.org/10.1007/978-94-007-7365-3

IFPRI (2014) Agricultural Science and Technology Indicators (ASTI). IFPRI, Washington. Available at https://asti.cgiar.org/publications/overview-publications. Accessed 1 July 2021

Karembu M, Nguthi F, Ismail H (2009) Biotech crops in Africa: the final frontier. ISAAA AfriCenter, Nairobi

Karembu M, Wafula D, Waithaka M, Belay G (2010) Status of biotechnology policies and biosafety legislation in the COMESA Region. ISAAA AfriCenter, Nairobi

Kochhar HPS, Gifford GA, Kahn S (2005) Regulatory and biosafety issues in relation to transgenic animals in food and agriculture, feeds containing genetically modified organisms (GMO) and veterinary biologics. In: Makkar HP, Viljoen GJ (eds) Applications of gene-based technologies for improving animal production and health in developing countries. Springer, Dordrecht, pp 479–498

Kuhlmann K (2015) Harmonizing regional seed regulations in sub-Saharan Africa: a comparative assessment. Written for the Syngenta Foundation for Sustainable Agriculture. Available via Syngenta. https://www.syngentafoundation.org/sites/g/files/zhg576/f/seedpolicy_new_africa_regulation_comparative_analysis_september_2015.pdf

Land O'Lakes International Development (2015) Public-Private Partnership for Artificial Insemination Delivery (PAID) 2015–2020. Available via Land O'Lakes International Development. https://www.landolakes.org/Where-We-Work/Africa/Tanzania/Public-Private-Partnership-for-Artificial-Insemina. Accessed 1 July 2021

Makinde D, Mumba L, Ambali A (2009) Status of biotechnology in Africa: challenges and opportunities. Asian Biotechnol Dev Rev 11(3):1–10

Mink SD (2016) Findings across agricultural expenditure reviews in African countries. IFPRI discussion paper 01522. IFPRI, Washington, DC. Available at https://media.africaportal.org/documents/130490.pdf. Accessed 1 July 2021

MLD (2006) National livestock policy. The United Republic of Tanzania, Ministry of Livestock Development. Dar-es-salaam. Available at https://www.tnrf.org/files/E-INFO_National_Livetock_Policy_Final_as_per_Cabinet_Dec-2006.pdf. Accessed 30 June 2021

MLFD (2015) Tanzania livestock modernization initiative, Dar-es-salaam: The United Republic of Tanzania, Ministry of Livestock and Fisheries Development. Available at http://livestocklivelihoodsandhealth.org/wp-content/uploads/2015/07/Tanzania_Livestock_Modernization_Initiative_July_2015.pdf. Accessed 30 June 2021

Mugabe J (2002) Biotechnology in sub-Saharan Africa: Towards a policy research agenda. ATPS special series paper 3. Africa Technology Policy Studies Network, Nairobi

Muraguri L (2010) Unplugged! An analysis of agricultural biotechnology PPPs in Kenya. J Int Dev 22:289–307

Nin-Pratt A (2016) Comparing apples to apples: a new indicator of research and development investment intensity in agriculture (September 23, 2016). IFPRI Discussion Paper 1559. Available via SSRN. https://ssrn.com/abstract=2845425. Accessed 1 July 2021

Olembo N, M'mboyi F, Oyugi K et al (2010) Status of crop biotechnology in sub-Saharan Africa. African Biotechnology Stakeholders Forum, Nairobi

Pray CE, Fuglie KO, Johnson DK (2007) Private agricultural research. In: Evenson R, Pingali P (eds) Handbook of agricultural economics, vol 3. Elsevier, London, pp 2605–2640

Ruttan VW (2000) Technology, growth, and development: an induced innovation perspective. Oxford University Press, Oxford

Shapiro BI, Gebru G, Desta S, Negassa A, Negussie K, Aboset G, Mechal H (2015) Ethiopia livestock master plan. Roadmaps for growth and transformation. A contribution to the growth and transformation plan II (2015-2020). Ministry of Agriculture, Livestock Resources Development Sector, Addis Ababa, Ethiopia

UNEP (2000) Initial strategy for assisting countries to prepare for entry into force of the Cartagena Protocol on Biosafety. In: United Nations Environment Programme (UNEP), Nairobi

UNEP (2006) Building biosafety capacity: the role of UNEP and the biosafety unit. UNEP-GEF biosafety unit, Geneva. Available at https://wedocs.unep.org/handle/20.500.11822/10000. Accessed 30 June 2021

UNEP (2017) National biosafety framework. Available via United Nations Environment Programme, Environment for Development. http://staging.unep.org/biosafety/National%20 Biosafety%20frameworks.aspx. Accessed 30 June 2021

Wafula D, Waithaka M, Komen J, Karembu M (2012) Biosafety legislation and biotechnology development gains momentum in Africa. GM Crops Food 3:72–77

World Bank (2016) News: Cameroon 100 million to boost livestock sector for improved productivity and climate change resilience. Available via World Bank. http://www.worldbank.org/en/news/press-release/2016/10/27/cameroon-100-million-to-boost-livestock-sector-for-improved-productivity-and-climate-change-resilience. Accessed 30 June 2021

The State of Applications and Impacts of Biotechnology in the Crop Sector

4

J. E. O. Rege, Dionysious Kiambi, Joel W. Ochieng, Charles Midega, and Keith Sones

Abstract

Application of biotechnology in the crop sector in sub-Saharan African countries was considered based on both the extent (e.g. limited use in laboratories versus widespread field use) and context (e.g. cloning individual genes versus cloning a whole organism) of use. Biotechnology types were allocated to three groups—low-, medium- and high-tech—depending on their relative complexity. High-tech applications were only encountered in a few countries. Consequently, low- and medium-tech applications were more important for categorizing most countries. Only one country, South Africa, was categorized as having 'very high' use of biotechnology in the crop sector. Burkina Faso, Ethiopia, Kenya, Madagascar, Malawi, Nigeria, Sudan, Tanzania, Uganda and Zimbabwe were categorized as 'high' use, and Benin, Botswana, Cameroon, Côte d'Ivoire, Eswatini, Ghana, Lesotho, Mali, Mauritius, Mozambique, Namibia, Senegal, Sierra Leone and Zambia as 'medium'. All other countries for which information was available were categorized as 'low' or 'very low' users of biotechnology in the crop sector. The crop sector has been the main beneficiary of initiatives to catalyse

J. E. O. Rege (✉)
Emerge Centre for Innovations-Africa, Nairobi, Kenya
e-mail: ed.rege@emerge-africa.org

D. Kiambi
African Biodiversity Conservation and Innovations Centre, Nairobi, Kenya

J. W. Ochieng
Agricultural Biotechnology Programme, University of Nairobi, Nairobi, Kenya

C. Midega
Poverty and Health Integrated Solutions (PHIS), Kisumu, Kenya

K. Sones
Keith Sones Associates, Banbury, UK

© The Author(s), under exclusive license to Springer Nature Switzerland AG 2022
J. E. O. Rege, K. Sones (eds.), *Agricultural Biotechnology in Sub-Saharan Africa*,
https://doi.org/10.1007/978-3-031-04349-9_4

biotechnology research and application, with the CGIAR's programmes being prominent in influencing applications.

4.1   Crop Agriculture in Sub-Saharan Africa

More than 60% of the population of SSA are smallholder farmers, and about 23% of the region's gross domestic product (GDP) comes from agriculture. Yet, SSA's full agricultural potential remains untapped; the region could produce two to three times more cereals and grains, which would add 20% more to the current worldwide 2.8 billion tons of output (FAO 2021). Similar increases are possible in the production of horticultural crops and livestock.

The major SSA staple food crops include maize, cassava, sorghum, millet, rice, yam, sweet potato, beans and groundnuts. In addition, SSA agriculture includes a large number of vegetables, both indigenous and exotic. Major export crops include coffee, tea, cotton, cocoa, flowers and fruit, among others. However, in many communities, major staples are grown as both food and cash crop; when a market is available, the major staple crop will be sold as a cash source for the household. Food self-sufficiency has been a food policy objective of many African governments, often narrowly defined to mean adequacy of food supplies of locally produced staple cereals (e.g. maize, sorghum, rice) to meet the country's per caput energy requirements.

Due to insufficient domestic production, SSA countries jointly spend about USD30 billion to USD50 billion a year to import food. Considering the size of the economies of these countries and the many development priorities they are unable to pursue, such as investment in infrastructure and social and economic amenities, this high food bill is a major challenge. Moreover, it is estimated that, if domestic production does not increase dramatically, Africa is likely to spend as much as USD150 billion on food imports by 2030.

In contrast to the constraints imposed by the sharply contrasting summer and winter seasons in most developed countries, many crops can be grown throughout the year in large parts of SSA as the continent is endowed with a tropical climate. Consequently, the increase in consumer demand in developed countries for out-of-season fresh fruit and vegetables is a major opportunity for African countries to produce these crops for export during the 'off-seasons' at attractive prices, so long as the required quality standards can be met. Moreover, the high labour cost in developed countries has made in-season production also an attractive proposition for Africa. In addition, high altitude regions of some African countries (e.g. Ethiopia, Kenya and Uganda in East Africa) have suitable climate for growing cool season crops requiring mild temperatures. Thus, rising costs of greenhouse heating, high labour costs and competitiveness in an increasingly global economy have made tropical countries a preferred alternative for producing some greenhouse crops. Some tropical fruits and vegetables also increasingly show substantial export promise as consumers' desire for variety and awareness of the health benefits of these crops increases. Vegetables exported from Africa include asparagus, snow peas, fine beans, round beans, baby carrots, baby corn, hard-shell garden peas, Brussels sprouts, broccoli, chillies and globe artichoke. Avocado, mango, passion fruit and pineapple make up the bulk of the fruit export. South Africa, Cote d'Ivoire and Kenya have historically been the leaders in the exports of these commodities; Zambia and Zimbabwe have more recently begun exporting. There has also been export to Europe (especially the UK) of substantial amounts of 'Asian vegetables' grown in Africa, e.g. Kenya. In eastern Africa, Kenya has been a leading exporter of flowers (roses and others). Ethiopia has recently joined and is said to be making major inroads in the lucrative flower business.

Africa's population is growing fast and so are the continent's food requirements. More has to be done to increase agricultural productivity than just expanding land use. The literature is awash with references to SSA as having large areas of untapped agricultural land that could be used to increase production of both staple and export crops, with estimates ranging from 480 million hectares to 840 million hectares. The McKinsey group has recently used estimates which consider dimensions such as market access, population density and agroecological conditions and concluded that only about 20–30 million hectares of additional cropland in SSA (representing 10% of additional cultivated land) is readily cultivatable today, but the area could increase as new infrastructure is put in place (Goedde et al. 2019). Moreover, while Africa continues to be highly targeted for large agricultural land deals, few of the deals have gone into implementation, and this, concludes the McKinsey team, suggests that land expansion will not be a major factor in increased production. Climate change is likely to exacerbate matters further. For instance, the yields from rainfed agriculture are set to decline in some countries and farmers need adaptation strategies, including drought-resistant crops and technologies and tools for ecological restoration.

There is a rich and diverse treasure trove of possible technological innovations. Biosciences, including modern biotechnology, is one area that provides a set of tools with which SSA can address its agricultural production and productivity challenges, including adaptation to climate change, and harness the many emerging opportunities—within the continent and beyond.

4.2   Application of Biotechnologies in the Crop Sector

The opportunities and challenges which characterize the crop improvement space in SSA include a focus on only a few mainly staple crops and inadequate research attention to the many crops that are central to nutrition security of communities (the so-called orphan or neglected crops); it takes up to 20 years for varieties to be replaced by new and improved ones; there is acute shortage of technical expertise and inadequate attention to succession planning that considers emerging technical skill areas and corresponding investment requirements; and overdependence by many countries on technologies developed by international agriculture research and development institutions, especially the CGIAR.

Areas of crop research that require attention in SSA and where biotechnology has a role to play include conservation action that facilitates continued access to clean (disease-free) and genetically appropriate (locally adapted) material for breeding, pre-breeding identification of novel genetic traits relevant for emerging challenges (e.g. climate change) and opportunities (markets), gene mining and trait introgression and actions to shorten the breeding cycle.

This section reviews the status of application of crop biotechnologies in SSA, including the relative use level by countries on a scale from 'very low' to 'very high'. In addition, a few case studies are presented (see Sect. 4.5) to illustrate how biotechnologies are being used and the impacts they are generating or could potentially generate.

4.3 Relative Level of Application of Crop Biotechnologies

The technology types under the three technology clusters (low-, medium- and high-tech applications) used in the analyses to determine the relative levels of application of biotechnology applications in crops in SSA countries are shown in Table 4.1.

4.3.1 Low-Tech Applications

The biotechnologies classified in this category are varietal development, bio-fertilizers, biopesticides, 'traditional' aflatoxin control technologies and vegetative propagation. Almost all SSA countries are involved in research and/or application of these low-tech applications and in the development and/or use of improved varieties developed through conventional breeding and/or varietal development.

Varietal development in Africa, supported by many international players, is one of the areas of crop biotechnology in which there has been substantial progress over the last three decades or more, but especially in the last two. The CGIAR is undoubtedly the most significant player in this domain. For example, the International Maize and Wheat Improvement Center (CIMMYT), together with the national agricultural research systems (NARS), is credited with a 23% share of improved maize variety releases in eastern and southern Africa; the corresponding figure for the CGIAR's contribution in West and Central Africa is 74% (Walker et al. 2014) (Table 4.2). The private sector is also well established in some countries, notably

Table 4.1 Crop technologies on the basis of which countries have been classified

Low-tech	Medium-tech	High-tech
• Classical crop breeding and varietal development based on selection and crossing • Bio-fertilizers • Biopesticides • Aflatoxin control • Vegetative propagation	• Tissue culture • Polymerase chain reaction (PCR) • Marker-assisted selection (MAS) • Bioinformatics	• Genetic modification • Gene editing techniques

Table 4.2 The contribution of the CGIAR to varietal output in SSA, 1980–2011

Crop	Number of released varieties related to CGIAR activity	Share of CGIAR-related varieties to total releases (%)[a]
Chickpea	23	95.8
Lentil	13	86.7
Pigeon pea	14	82.4
Potato	72	75.0
Yam	26	74.3
Maize-WCA[b]	173	74.2
Cassava	143	68.1
Sweet potato	59	66.3
Cowpea	88	57.5
Rice	179	51.4
Soybean	69	48.9
Wheat[c]	81	45.0
Groundnut	41	43.6
Pearl millet	45	40.2
Faba bean	10	40.0
Bean	88	39.1
Sorghum	38	24.8
Maize-ESA[c]	171	22.8
Barley	8	21.1
Banana	1	16.7
Field pea	4	16.7

[a]The share estimate is understated because data collected in smaller producing countries did not contain information on the institutional source of genetic material since 2000
[b]WCA = West and Central Africa
[c]ESA = East and southern Africa
Source: Walker et al. (2014)

South Africa, Kenya, Zambia and Zimbabwe, where hybrid maize varieties dominate the market.

Since 2006, CIMMYT has released more than 100 varieties in SSA as part of its Drought Tolerance for Maize in Africa Initiative. Overall, the contribution of CGIAR research centres to improved varieties ranges from about 17% for field pea to 96% for chickpea. The French Agricultural Research Centre for International Development (CIRAD) has also played a significant role in generating materials that have resulted in varietal change in several food crops in West Africa, including roots, tubers and bananas.

Mycorrhizal-based bio-fertilizers have a long history of use in SSA. The technology was first promoted by the Microbial Resources Centre Network that was started with UNESCO funding in the 1970s (UNDP and UNESCO 1994). Several SSA countries, including Kenya, Malawi, Senegal, Tanzania, Uganda, Zambia and Zimbabwe, were members of the network. One example of a bio-fertilizer technology is Biofix, a product used with legume crops which is based on nitrogen-fixing bacteria, that was developed by Nairobi University in Kenya and has been commercialized and adopted widely in SSA (UNDP and UNESCO 1994). Other examples of the use of this technology in West and Central Africa include production of rhizobial-based bio-fertilizers in DRC, production of mycorrhizal-based bio-fertilizers for rural markets in Gabon and application of rhizobium inoculum and bio-fertilizer for commercial production of soya beans in Nigeria (Olembo et al. 2010; Alhassan 2001). There are other examples of the use of the technology in eastern and southern Africa: in Zimbabwe there is a legume inoculant factory developing rhizobium strains for farmers. In Malawi, the Chitedze Agricultural Research Station is involved in bio-fertilizer production and biological control in pest management research, while the University of Burundi is undertaking research and application in bio-fertilizers and also mycoculture of mushrooms. In Sudan, the Agricultural Research Corporation is involved in the production of rhizobial-based bio-fertilizers and *Bt*-based bioinsecticide. However, despite the widespread use of this technology, there are still several countries, such as Djibouti, Guinea Bissau, Liberia, Sierra Leone and Somalia, where use of bio-fertilizers and biopesticides has been limited or non-existent.

Across SSA there are numerous public and private partnerships supporting research and application of technologies for the prevention and control of aflatoxin contamination. The activities include research and analysis, development of new control technologies and scaling up existing technologies. For example, with support from the USAID-supported Africa Research in Sustainable Intensification for the Next Generation (Africa RISING), a management strategy using biocontrol products containing native atoxigenic *Aspergillus flavus* fungi to reduce crop aflatoxin content has been developed by the International Institute of Tropical Agriculture (IITA) and partners. The products, marketed as Aflasafe™, each containing four atoxigenic *A. flavus* strains, have been tested in maize and groundnut farmers' fields in efficacy trials in several countries. To date these products have been registered with pesticide regulatory authorities for use in ten African nations (see impact case study in Sect. 4.5.5): Burkina Faso, Ghana, Kenya, Malawi, Mozambique, Nigeria, Senegal, Tanzania, the Gambia and Zambia. In 2021, products are being developed for another ten countries (Konlambigue et al. 2020).

In addition, the Feed the Future Peanut and Mycotoxin Innovation Lab is conducting some advanced research on the use of RNA interference to reduce aflatoxin in peanut seeds through genetic diversity of aflatoxigenic *Aspergillus* species and genetic transformation of peanut plants. This work is being undertaken in collaboration with NARS in Kenya, Malawi, Mozambique, Uganda and Zambia. Other ongoing aflatoxin research activities include establishment and validation methods for measuring major mycotoxin biomarkers, especially the aflatoxin-lysine adduct, in human dried blood spot samples to support urgent needs of nutritional studies and testing of aflatoxin levels in food crops in many African countries, including Burkina Faso, Ghana, Kenya, Malawi, Mozambique, Rwanda, Tanzania, Uganda, Zambia and Zimbabwe (Bandyopadhyay et al. 2016).

4.3.2 Medium-Tech Applications

Medium-tech applications are made up of two main subcategories: tissue culture and molecular marker applications. The other types of medium-tech applications that are widely used in SSA are gene banking, mutation breeding and biofortification.

Tissue Culture   Tissue culture or in vitro culture is a technique through which small plant organs, embryos, cells or protoplasts are aseptically isolated and grown on artificial nutrient media under controlled environment into complete plants. Since the inception of tissue culture in the twentieth century, the technology has been widely used in crop production in Africa, particularly in germplasm conservation, micropropagation and production of pathogen-free plants, embryo rescue and genetic transformation. Tissue culture has been applied widely across SSA in many food crops (Olembo et al. 2010; Kyei et al. 2017).

In eastern Africa, tissue culture techniques, especially micropropagation and production of pathogen-free materials, are widely applied in most of the government research institutions and public universities. In Kenya, tissue culture has been used to develop superior banana genotypes that have been widely cultivated in Kenya and Uganda (Masiga et al. 2013; IFPRI 2014). The techniques have been used by the Kenya Agricultural and Livestock Research Organization (KALRO) in micropropagation, disease elimination and genetic transformation in macadamia, sweet potatoes, pyrethrum, cassava, Irish potatoes, oil palm and vanilla. It has also been used in generation of double haploids in wheat. Private companies and non-governmental organizations are involved in application of tissue culture techniques for the development of elite germplasm of crops including cash crops, staples and vegetables.

In Tanzania, tissue culture has been employed for mass propagation of virus-free planting material, conservation of vegetatively propagated crops, genetic transformation and embryo rescue in coconut, sisal, cassava, sweet potato and banana, among other crops. The technology is being applied in most of the agricultural research institutions and public universities (Virgin et al. 2007; Mtui 2011; Masiga et al. 2013). In the other eastern African countries—Burundi, Ethiopia, Sudan and Uganda—tissue culture technologies have equally been applied in mass propagation, disease elimination, embryo rescue and genetic transformation in a number of crops (Mtui 2011; Kyei et al. 2017).

In West and Central Africa, tissue culture technologies are widely applied in crop research and development. The Institute of Agricultural Research in Nigeria routinely uses tissue culture in micropropagation and in vitro conservation of cassava, yams and banana, as well as embryo rescue in yams, while in Cameroon, the Institute of Agricultural Research and Development has been using tissue culture in the propagation of cocoa, yams, coffee, banana, oil palm and cotton (FARA 2011). In DRC, for example, the University of Kisangani is using tissue culture for in vitro propagation of potatoes, soybean, maize and rice, while in Niger, the National Institute of Agricultural Research is using tissue culture in the development and multiplication of disease-free cassava materials.

In southern Africa, the national agricultural research institutions and universities regularly use tissue culture techniques in micropropagation, in vitro conservation, disease elimination and somatic hybridization of a wide range of crops including cassava, potato and bananas (Moyo et al. 2011; Olembo et al. 2010; Kyei et al. 2017).

There was no recorded use of tissue culture in most of the countries classified in 'low' and 'very low' biotechnology user categories (Table 3.3) which include Angola, Djibouti, Equatorial Guinea, Gabon, Liberia, Somalia and South Sudan, among others.

Biofortification Biofortification is the process of breeding nutrients into staple food crops to help reduce mineral and vitamin deficiencies. Biofortification offers a cost-effective and sustainable investment compared to supplementation and commercial fortification of foods. This is because it is a one-time investment in plant breeding that yields micronutrient-rich planting materials for farmers to grow for years to come. There are three common approaches to biofortification—agronomic, conventional breeding and transgenic—and these can be classified as low-, medium- and high-tech applications, respectively. Agronomic biofortification provides temporary micronutrient increases through fertilizers and/or foliar sprays and is useful to increase micronutrients absorbed directly by the plant. Conventional plant breeding involves identifying and developing parent lines with naturally high vitamin or mineral levels and crossing them over several generations to produce plants with desired nutrient and agronomic traits. Transgenic plant breeding seeks to do the same in crops where the target nutrient does not naturally exist, by inserting genes from another species. A three-stage process takes transgenic biofortification strategies from discovery through development to delivery (see Fig. 4.1, Source: Bouis et al. (2017)).

Starting in 2012, significant progress has been made with biofortification in several SSA countries (Bouis et al. 2017). A biofortification case study is presented in Sect. 4.5.6. With strong proof-of-concept for biofortification, moving towards scale will require increased public and private sector investment in crop development and seed systems to sustain the pipeline of biofortified varieties.

Molecular Markers Molecular markers are fragments of DNA that can be easily tracked and quantified in a population that may be associated with a particular gene or trait of interest. The use of molecular markers in the development of improved varieties has been increasing consistently in SSA over the last two decades. This includes their application in genetic diversity studies, genomics, genotyping,

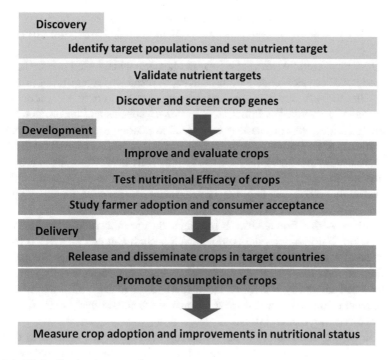

Fig. 4.1 Biofortification impact pathway

association mapping, quantitative trait loci (QTL) mapping and marker-assisted selection (MAS).

Molecular markers are an important tool for MAS which only requires young seedling stages to tag and introgress genes, thus enabling pyramiding of genes and selection of QTLs simultaneously (Brink et al. 1998; Edwards and McCouch 2007). The genetic markers in use range from first generation markers, such as amplified fragment length polymorphisms, random amplified polymorphic DNA markers (RAPDs) and isozymes, to more technologically advanced markers such as simple sequence repeats (SSRs), single nucleotide polymorphisms (SNPs) and diversity arrays technology (DarTs) (Kyei et al. 2017).

In eastern Africa, the technology is widely used across many crops. For example, in Kenya, marker technology has been used for characterization and mapping of maize streak virus and grey leaf spot resistance genes in maize; development of drought-tolerant maize and wheat; development of wheat lines resistant to the Russian wheat aphid; development of *Striga*-resistant sorghum; selection for smut resistance in sugarcane; diversity studies for sweet potato, cassava, sorghum, tea and coffee; and characterization of indigenous species of forages, among others.

In Tanzania, molecular markers have been used extensively in MAS in cassava and farmer participatory improvement of cassava germplasm for farmer- and market-preferred traits (Virgin et al. 2007). They have also been used in characterization and diversity studies of cashew germplasm by the Cashew Biotechnology Unit at Mikocheni Agricultural Research Institute (Olembo et al. 2010; Mtui 2011; IFPRI 2014).

In West and Central Africa, molecular markers are widely used in crop improvement programmes. For example, in Ghana, molecular fingerprinting has been applied in various crops including cassava, yam, coconut, frafra potato (*Solenostemon rotundifolius* Poir.), banana (*Musa*) species and cocoa. Researchers at CSIR have used genetic markers and MAS in genotyping, characterization and improvement of groundnuts, coconuts and plantain (Quain and Asibuo 2009). In Nigeria, molecular markers have been used extensively in MAS for the improvement of maize and cassava; DNA fingerprinting of cassava, yams and banana and pests and microbial pathogens; and the development of genome linkage maps for cowpeas, cassava, yams and banana (Kyei et al. 2017; Alhassan 2001).

In southern Africa, South Africa is leading in the use of molecular markers in crop improvement. For example, the Agriculture Research Council's (ARC's) biotechnology laboratory routinely uses molecular markers in diagnostics for pathogen detection; cultivar identification in potatoes, sweet potato, ornamentals, cereals and cassava; seed-lot purity testing in cereals; MAS in maize and tomato; and the development of markers for disease resistance in wheat and many other food crops (Mulder 2003; Moyo et al. 2011).

In Zimbabwe, the Biotechnology Research Institute is using MAS to develop maize varieties that are drought-tolerant and resistant to pests. As an initial step, QTL underlying drought tolerance and insect resistance were identified in the local population. DNA markers are routinely used in MAS for drought tolerance and insect resistance in maize, cultivar identification, gene tagging and diversity studies in maize, sorghum, pearl/finger millet, wheat, barley and tobacco. In the University of Zimbabwe, research work has been conducted to identify molecular markers that are linked to witchweed (*Striga asiatica*) resistance for eventual use in MAS. The occurrence of interspecific and intraspecific strains of *S. asiatica* has been investigated using markers (Olembo et al. 2010; Kyei et al. 2017). Moreover, all the SSA countries in the 'medium' to 'very high' use categories (Table 2.5) are routinely using molecular markers, to different extents, for research and development applications in the generation of improved crop varieties (Kyei et al. 2017). There is no recorded information of the use of molecular markers in 'low' or 'very low' use countries in Table 2.5.

4.3.3   High-Tech Applications

The high-tech applications have primarily focused on enhancing agricultural productivity using genetic engineering—a technique of controlled manipulation of genes to change the genetic makeup of cells and to move genes across species boundaries to produce novel organisms, that is, to produce genetically modified organisms (GMOs).

Sub-Saharan Africa is regarded as the region with the biggest potential to benefit from adoption of genetically modified (GM) crops. This is because of the magnitude of poverty and malnutrition in the region. In 2018, only three countries (Eswatini, South Africa and Sudan) were commercially pursuing GM crops. An additional three countries (Malawi, Nigeria and Ethiopia) approved the use of GM crops in 2019, and Kenya witnessed approval of commercial cultivation of *Bt* cotton and formal application by Kenya Agriculture and Livestock Research Organization (KALRO) seeking government approval for open field cultivation of cassava brown streak disease (CBSD)-resistant cassava. The approval in 2019 by Nigeria of the insect protected (*Bt*) cowpea that is resistant to pod borers which can cause up to 80% yield loss was a major landmark: Nigeria became the first country ever to commercialize genetically improved cowpea globally. At the same time, the International Institute of Tropical Agriculture (IITA) submitted an application to Nigeria's National Biosafety Management Agency for confined field trial of genetically engineered cassava with increased starch yield. In total, *Bt* cotton was grown on more than 1,000 farms in Nigeria as part of large-scale trials preceding planned scaling.

Although not formally approved for commercial application, significant traction is visible in GM crop research and regulation processes in Mozambique, Niger, Ghana, Rwanda and Zambia in 2020. Cameroon, Ghana, Tanzania and Uganda are also trialling various strains of GM crops as the first step in the approval process. In 2018, Eswatini was the 26th highest ranking country globally by area of GM crop grown, all of it cotton.

South Africa ranked ninth globally in terms of crop land under GM crops in 2018. The country has grown GM crops for more than 20 years and leads in their adoption on the African continent with plantings of GM maize, soybean and cotton totalling over 2.7 million hectares in 2017 (ISAAA 2015b, 2016; USDA 2016).

Earlier on, Burkina Faso was the second SSA country to commercialize GM crops. By 2013 the country had 470,000 hectares under *Bt* cotton with a 68% adoption rate, and the country was also conducting national performance trials for *Bt* cowpea and approvals for confined field trials for *Bt* maize had been granted (ISAAA 2015a). However, in 2015 the country abandoned growing *Bt* cotton as buyers complained the fibres were of poor quality (see Sect. 4.5.2).

Sudan approved its first GM crop, insect-resistant *Bt* cotton, for commercial planting in 2012 and has gradually increased the area under this crop to 192,000 hectares in 2017. The rate of adoption of biotech cotton is high at 98% and only a few farmers grow non-*Bt* cotton (ISAAA 2016). In 2019, Sudan ranked 15th in terms of area planted with GM by country.

There are clear disparities in the development and adoption of GM technologies for the other SSA countries (Midling 2011; Bailey et al. 2014; Chambers et al. 2014). In eastern Africa, there are several advances in the development and adoption of GM technologies. Some of the activities include confined field trials for *Bt* (insect resistance) and drought tolerance for banana, cassava, cotton maize and sweet potato in Ethiopia, Kenya, Tanzania and Uganda. There are no GM technology development and deployment activities reported in Burundi, Rwanda or Eritrea.

In West and Central Africa, other than the cases of Burkina Faso and Nigeria referred to above, other GM-related activities include contained field trial for *Bt* cotton, *Bt* maize, *Bt* cowpeas, NUWEST rice and biofortified sorghum in Cameroon, Côte d'Ivoire, Ghana, Nigeria, Senegal and Mali.

Table 4.3 presents a summary of status of development and application of GM technology for crops in SSA.

Figure 4.2 summarizes status of research and use of biotechnologies by regions of Africa.

Generally, the focus on GM technology is on traits of high relevance to the major challenges facing the continent—pests, diseases and drought. In terms of insect-resistant crops, *Bt* cotton remains the most widely used across SSA with Eswatini, Ethiopia, Malawi, Nigeria and Sudan having planted it on close to 244,000 hectares in 2019. In the same year, South Africa planted *Bt* maize on approximately 175,000 hectares. A total of approximately 694,000 hectares of herbicide-tolerant soybean was planted in South Africa, similar to the acreage in 2018. Herbicide-tolerant cotton was grown on 2183 hectares. Out of the 1.95 million hectares of GM maize planted in South Africa in 2019, 58% was stacked (approximately 1.13 million hectares), that is, they had more than one genetic modification. *Bt* herbicide-tolerant cotton was grown on approximately 41,471 hectares. The year 2019 also marked South Africa's 22nd year of commercial cultivation of the three principal biotech crops: cotton, maize and soybean.

From 2008 to 2018, AATF coordinated a public-private partnership called Water Efficient Maize for Africa (WEMA) to develop drought-tolerant and insect-protected maize using conventional breeding, marker-assisted breeding and other biotechnology, with a goal to make these varieties available royalty-free to smallholder farmers in SSA through African seed companies. The projects, featured in more detail in Sect. 4.5, illustrates the status of research and application of some high-tech biotechnologies in SSA delivered through public-private partnerships (e.g., the AflasafeTM).

4.4   Classification of Countries

As shown in Table 4.4, SSA countries are at widely differing stages in research and application of biotechnology in crops. The majority of the countries are involved in both research and use primarily of low-tech applications, such as bio-fertilizers, biopesticides and a wide range of tissue culture techniques. Some have made advances and are applying medium-tech, such as MAS, and also high-tech, such as genetic engineering (ISAAA 2016).

On the basis of this analysis, SSA countries have been classified in the five biotechnology application categories (Table 4.4). Among countries with information available, only one (South Africa) is ranked as 'very high' in application, while the 'high use' category has ten countries, namely, Burkina Faso, Ethiopia, Kenya, Madagascar, Malawi, Nigeria, Sudan, Tanzania, Uganda and Zimbabwe. The rest of the countries are in the 'medium' (14), 'low' (16) and 'very low' use (3) categories.

Table 4.3 Status of genetic engineering technology research and applications in SSA

Crop	Burkina Faso	Ghana	Kenya	Malawi	Mozambique	Nigeria	South Africa	Sudan	Tanzania	Uganda	Zimbabwe
Bananas										CFT	
Canola							CR, CFT				
Cassava			CFT			CFT	TR			CFT	TR
Cotton	CR, CFT	CFT	CR	CR	~CFT		CR, CFT	CR		CFT	CFT
Cowpeas	CFT	CFT				CR					
Maize			NPT		~CFT		CR, CFT		~CFT	CFT	~CFT
Pigeon peas			TR/ GH								
Potatoes							TR				
Rice		CFT					TR			CFT	TR
Sorghum	CFT		CFT			CFT	TR				
Soybeans							CR, CFT				
Sugarcane							CR, CFT				
Sweet potatoes		GH	CFT							GH	
Tobacco									CFT		
Wheat							CFT				

Notes: *CFT* confined field trials; *CR* commercial release; *GH* greenhouses; *TR* transformations; *NPT* national performance trials

Source: Chambers et al. (2014), Kiome (2015), ISAAA (2016)

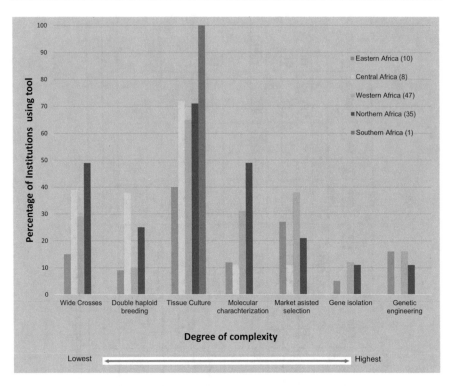

Fig. 4.2 Biotechnology tools researched and applied in Africa. *Source: Compiled from FAO-GIPB (2011) and Chambers et al. (2014). Notes: the number in parentheses beside each region represents the number of institutions surveyed in the region. (1) Eastern Africa includes data from Eritrea, Ethiopia, Kenya, Madagascar, Malawi, Mozambique, Rwanda, Uganda, Zambia and Zimbabwe; (2) Central Africa includes data from Angola, Cameroon and Gabon; (3) southern Africa includes data from Namibia only; (4) western Africa include data from Benin, Burkina Faso, Côte d'Ivoire, Ghana, Mali, Niger, Nigeria, Senegal, Sierra Leone and Togo; and (5) northern Africa includes data from Algeria, Morocco, Sudan and Tunisia (in this figure, Sudan refers to the former Sudan, which is now two independent nations, Sudan and South Sudan)*

Table 4.4 Classification of countries on the basis of applications of crop biotechnologies

Application category	Countries
Very low use	Equatorial Guinea, Guinea, Somalia
Low use	Angola, Burundi, CAR, Chad, Congo, Djibouti, DRC, Eritrea, Gabon, Gambia, Guinea Bissau, Liberia, Niger, Rwanda, South Sudan, Togo
Medium use	Benin, Botswana, Cameroon, Côte d'Ivoire, Eswatini, Ghana, Lesotho, Mali, Mauritius, Mozambique, Namibia, Senegal, Sierra Leone, Zambia
High use	Burkina Faso, Ethiopia, Kenya, Madagascar, Malawi, Nigeria, Sudan, Tanzania, Uganda, Zimbabwe
Very high use	South Africa

Countries not listed due to lack of sufficient data: Cape Verde, Comoros, Mauritania, Sao Tome and Principe and Seychelles

4.5   Impact Case Studies: Crops

Predictive economic studies conducted by African economists in collaboration with researchers from the International Food Policy Research Institute, in five countries across the region, show the total potential benefits of adopting biotech crops. A snapshot of impacts documented by ISAAA (2019) includes the estimated cumulative economic benefit between 1998 and 2018 by South Africa of USD2 billion from biotech crops and USD237 million for 2018 alone. Sudan reported positive change in the entire cotton sub-sector value chain, starting from production, through reduced number of chemical sprays and higher yields, with farmers reporting increased income as well as improved living and health standards, and, in Eswatini, farmers reported a reduction of chemical sprays from eight to just three per cropping season—a significant benefit to the environment.

Nine case studies are summarized in this section: Pan-African Bean Research Alliance (PABRA) as an example of an institutional arrangement for developing and scaling technologies relevant for smallholder farmers, *Bt* cotton in Burkina Faso, tissue culture banana in East Africa, Water Efficient Maize for Africa (WEMA), Aflasafe™ mitigating aflatoxin risks, making biofortification work for nutrition, maize lethal necrotic disease, biotechnology options for managing *Striga* in cereals in Africa and exploiting innate plant defences for control of maize stemborers.

4.5.1   The Pan-Africa Bean Research Alliance

The common bean (*Phaseolus vulgaris* L.) is an important source of protein and a micronutrient-rich food for people in Africa, especially in maize-dominated eastern and southern Africa. It is a valuable cash crop, especially for women, and plays an important role in farming and agroecosystems due to its ability to fix atmospheric nitrogen in cereal-dominated cropping systems. Production of common bean is, however, constrained by a wide range of challenges, such as pests and diseases, poor soil fertility and climate change, including erratic rainfall, drought, flooding, heat and cold stress, all of which significantly impact on yield.

Concerned by declining bean production and productivity while the demand was on the rise, the Pan-Africa Bean Research Alliance (PABRA) was established in 1996 with the aim to enhance food and nutrition security, income generation, poverty reduction, empowerment and health of poor communities through bean research, capacity building, networking and partnership building on beans.

Today, PABRA, facilitated by Bioversity International and CIAT, working together as the Alliance Bioversity-CIAT, is a mature model of a research and development partnership that brings together national agricultural research institutes of 31 countries across SSA, universities in these countries, the Alliance and more than 500 value chain actors (private and producers' organizations). They are grouped in three regional networks which have a joint planning framework to coordinate research implementation and achieve synergy in outcomes. Through nine bean corridors, PABRA has adopted a food systems approach that links consumption to responsive production systems, supported by demand-led bean research.

A key feature of PABRA was the shift away from the previously dominant breeding strategy among NARS in Africa, which focused on breeding for improved yield or resistance to a single environmental stress, to a food systems approach carried out by multidisciplinary teams using a demand-led research approach along bean value chains. The new approach was to design breeding programmes that addressed the demands of both farmers and consumers, including bean grain size, colour, cooking time and higher iron and zinc, as well as resilience to the major biotic and abiotic production constraints. A broadly based participatory approach was used which enabled a wide range of value chain actors to contribute to the development of breeding strategies and goals and varietal selection for each market class of bean.

Breeding programmes undertaken under the auspices of PABRA rely on conventional breeding, but increasingly advances in biotechnology are being utilized, especially MAS for disease resistance, genomic selection to optimize breeding for both consumer and agroecological traits and variety DNA fingerprinting. In addition to developing and releasing new farmer- and consumer-demanded varieties, PABRA also aims to ensure that appropriate seed is available to farmers through demand-led seed systems. Promising breeding lines or new bean varieties developed in one country are shared with other countries where they may be suitable. In this way, countries which lack the capacity to undertake their own breeding programmes can still benefit from the improved varieties.

For the last 25 years, PABRA's key achievements have been the development and release of more than 700 improved bean varieties with various traits such as beans rich in iron and zinc, climate-smart varieties that are early maturing, drought-tolerant and pest- and disease-resistant. More than 35 million farmers (58% of them women) have been reached by PABRA seed system actors (Farrow and Muthoni-Andriatsitohaina 2020). Average bean yields have increased by 36% or even doubled when the new varieties have been adopted in Burundi, Ethiopia and Uganda (Oratungye et al. 2021a, b; Katungi et al. 2020; Habte et al. 2021). This has generated an additional USD500–800 per hectare per season in rain-fed conditions. This increased production has translated into more bean business opportunities, opening up to over 250 small and medium enterprises which are commercializing various bean-based products and providing market opportunities for 4 million farmers, 51% of whom are women (Oratungye et al. 2021a). PABRA's lessons from sustainable food systems, demand-led breeding, responsive seed systems and multidisciplinary approaches are expanding to other legume crops, including pigeon pea, groundnut and soybean.

In 2019, PABRA's work has been recognized internationally when it was awarded the Al-Sumait Prize for its contribution in food security in Africa, and the PABRA model was nominated among the 'Golden eggs' under One CGIAR as a model of success on partnerships contributing to sustainable food systems.

4.5.2   Bt Cotton in Burkina Faso

Cotton is a very important and politically and economically sensitive crop in Burkina Faso. More than 300,000 smallholder farmers cultivate cotton, which is the number 1 cash crop in the country and contributes 25% of agricultural GDP. It also supplies over half the cooking oil used in the country and provides animal protein for the livestock industry as by-products of cotton fibre processing.

Cotton produced in Burkina Faso has long enjoyed an excellent reputation for quality and has been in high demand in international markets for decades, attracting premium prices. The high quality is due to a combination of more than 70 years of painstaking selective breeding and the fact that the crop is handpicked. This produces cotton with a highly desirable long fibre length and a high ginning ratio—that is, the percentage of fibre produced per unit weight of raw cotton delivered from the field. The local cultivars are also very well adapted to local conditions.

The cotton industry in Burkina Faso is highly organized and regulated and vertically integrated. The large cotton companies supply smallholders with inputs, such as seed, pesticides and fertilizer, on credit and buy back the harvest at a guaranteed price.

Cotton usually requires multiple cycles of insecticidal spray to control pests, especially caterpillars of various species of moth. Heavy use of these pesticides is both costly and hazardous. In addition, in the 1990s in Burkina Faso, some of the major pest species, including cotton bollworm, became resistant to the commonly used insecticides.

This situation prompted the Burkina Faso authorities to explore the use of a cotton variety genetically engineered for protection against pests known as *Bt* cotton. *Bt* cotton was introduced by Monsanto in the USA in 1995. It was produced by inserting genes that code for endotoxins from *Bacillus thuringiensis* (*Bt*) into cotton plants. These endotoxins, known as Cry proteins, are toxic to various species of insect larvae including the cotton bollworm, a type of moth.

In 2003, the Burkina Faso government signed an agreement with Monsanto for field trials of *Bt* cotton containing genes that coded for two Cry proteins, CryAc and Cry2Ab. The initial trials were reported to show *Bt* cotton had better resistance to the insect pests and gave higher yields compared to local cultivars. However, eager to maintain their reputation for quality cotton, the authorities did not want to import US germplasm; rather they proposed inserting the *Bt* trait into local cotton cultivars. Monsanto achieved this through three generations of backcrossing their *Bt* cotton with local cultivars to generate a stable commercial hybrid, sold under the trade name Bollgard II. This was jointly patented by Monsanto and the local cotton industry and released in 2009.

Due to the highly organized and centralized nature of the cotton industry in the country, it was relatively easy to rapidly substitute the *Bt* cotton seed for conventional seed. By 2013, more than 70% of the cotton acreage in Burkina Faso was *Bt* cotton.

The Bollgard II seed was 30 times more expensive than conventional seed, but savings in use of pesticides and yield increases of more than 20% observed in the trials resulted in an anticipated increase in profit of more than 50%. However, in the first years after commercialization, officials in Burkina Faso noticed that the fibre length of the *Bt* crop was shorter and the ginning ratio had declined compared to the conventional local cultivars. Burkina Faso's international reputation for high-quality cotton was quickly being undermined. While neighbouring countries growing conventional high-quality cotton continued to sell their cotton at premium prices, Burkina Faso's harvest went unsold.

The reason for the decline in fibre length is not clear. In theory, inserting the genes for the *Bt* trait should have had no effect on any other traits, but in practice this process clearly had a highly detrimental impact on quality traits too. Frustrated by Monsanto's inability to explain what had happened, much less to fix the problem, the cotton industry took action and rapidly phased out use of *Bt* cotton. In addition, they demanded USD280 million as compensation from Monsanto for losses incurred, although it is reported that a much lower figure was eventually agreed upon as compensation.

Not surprisingly, the negative experience in Burkina Faso has been seized upon by those opposed to the use of GM crops.

One of the suggestions for the root cause of the problem is that three generations of backcrossing were insufficient. Although this is standard practice in the USA, where the breeding objective is focused solely on pest control, some experts have suggested that five or more generations might be needed to ensure the carryover of the desirable traits linked to quality as well as adaptation to local conditions from the local cultivars. It has been suggested that in the desire to get *Bt* cotton seed to market, the interests of Monsanto and the long-term interests of the local cotton industry may not have been aligned. With each generation taking an additional year, adding two or more additional generations would have delayed commercialization by at least 2 years.

With the return to conventional cotton, reports suggest that farmers are now struggling to control bollworm and have had to significantly increase pesticide use. Some are keen that *Bt* cotton is reintroduced, but the local authorities have made it clear that before that could happen, the problem with cotton quality would have to be addressed. In 2018, the German chemical giant Bayer bought Monsanto, and so any future reintroduction of *Bt* cotton in Burkina Faso would likely be undertaken by the new company.

Sources: Bavier (2017), Dowd-Uribe and Schnurr (2016), Showalter et al. (2009), Traore (2016)

4.5.3 Tissue Culture Banana in East Africa

One of the problems affecting the production of banana as a reliable food and commercial crop is the high incidence of pests and diseases. Through traditional practices of exchanging banana suckers, farmers can transmit banana pests and diseases (Panama disease, black sigatoka, weevils and nematodes being among the most economically important) from plant to plant or farm to farm, and this can reduce yields by up to 90%.

Tissue culture (TC) technology, also known as micropropagation, has been successfully used to enhance farmers' access to disease-free planting materials since the 1960s. It does not involve genetic modification; rather the plants produced are all clones, genetically identical to the parent plants. The end product is similar to that produced by taking traditional cuttings from the parent plant, but TC enables many more identical plants to be produced under more controlled conditions.

The International Service for Acquisition of Agri-biotech Applications (ISAAA), in collaboration with the Kenya Agricultural Research Institute (KARI), initiated a TC banana project in 1996. Smallholder farmers were involved in the identification of varieties through participatory rural appraisal. An inclusive and interactive strategy for technology transfer, using the Farmer Field Schools' approach, with feedback and monitoring mechanisms, was applied to ensure ownership and sustainability.

Since 2003, Africa Harvest, an NGO headquartered in Nairobi, has promoted the widespread adoption of TC banana in Kenya, using a whole value chain approach that incorporates private companies and public institutions that provide plantlets to farmers. In addition, and as part of this collaboration, KARI (now KALRO) and Jomo Kenyatta University of Agriculture and Technology produced plantlets in their laboratories for distribution to TC nurseries managed by farmer groups in various parts of the country. A microcredit facility scheme was also incorporated.

The TC banana project made it possible for more than 10,000 Kenyan farmers to access large quantities of superior clean planting material with early maturity traits (12–16 months versus 2–3 years for conventional bananas), bigger bunch weights (30–45 kg vs. 10–15 kg) and higher annual yield (40–60 tons per hectare vs. 15–20 tons). After two decades, however, less than 10% of all banana farmers in the country had adopted TC banana (Njuguna et al. 2010), although the figure is slightly higher at 15% in Central and Eastern Kenya where most of the dissemination started (Kabunga et al. 2012). This low rate of adoption has been attributed to the high costs of TC plantlets and also the more demanding management and inputs required to achieve high yields (Qaim 1999).

Studies seeking to assess the impact of TC banana in Kenya since its introduction in 1996 have been equivocal. While Muyanga (2009) reported insignificant yield differences, Kabunga et al. (2012) found a moderate productivity gain associated with TC adoption. Meanwhile, Njuguna et al. (2010) reported large productivity gains. These disparities in impact estimates have been attributed to temporal factors and water availability (Kabunga et al. 2012).

Farmers who adopted TC banana were also advised to adopt improved management practices and higher input use. Tissue culture banana can therefore be viewed as a technology package, the successful adoption of which requires not just access to TC banana planting material but also knowledge about how to grow them and willingness and ability to purchase additional inputs, such as fertilizer and irrigation. Lack of these could limit the benefits of TC technology, as exemplified by an assessment of adoption and impact of TC banana in Burundi, which, just like the Kenyan case, showed unexpected results—nonsignificant causal effects on banana yield and gross margin differentials between TC banana adopters and non-adopters (Ouma et al. 2013). The lack of significant impact was attributed to a different institutional setting in Burundi for promoting TC bananas compared to other countries, especially the use of poor-quality TC plantlets in the field and poor field management practices by smallholders in the Burundi setting.

Similarly, a study in Central Uganda (Mulugo et al. 2020), in an area where TC banana planting materials had been promoted for more than a decade, where there was a high prevalence of banana diseases that constrained conventional production, where TC banana nurseries had been established close by and where the farmers had been trained on how to grow and manage TC banana, found that just a third of farmers had adopted the technology. Women were more likely to adopt TC bananas than men.

New business and employment opportunities have sprouted along the banana value chain, particularly at the levels of production, distribution and postharvest levels, in Kenya and neighbouring countries where the planting of TC banana has been replicated.

Lessons learned through the implementation of the project included the following: participatory approaches in problem identification and technology delivery can increase probability and levels of success; involving the target beneficiaries in introducing new technologies helps ensure information, knowledge and skills are availed to farmers in a most cost-effective way that promotes uptake; public-private partnerships and complementarity are critical in the dissemination of technologies to resource-constrained farmers; quality end-user engagement to determine preferred technologies (banana varieties in this case) enhances technology uptake; and linking production to value addition and marketing is crucial for farmers to maximize benefits from an agricultural technology.

Sources: Wambugu and Kiome (2001), Wakhusama and Mbogo (2003), Karembu (2007), Nguthi (2009), Kikulwe et al. (2012)

4.5.4   Water-Efficient Maize for Africa (WEMA)

Maize is the staple food for more than 300 million people in Africa. The major production constraints for this important food security and cash crop are drought and vulnerability to stemborers, which are the larvae of some moth species particularly *Chilo partellus* and *Busseola fusca*. In Kenya alone, stemborers reduce maize production by a yearly average of 13% or 400,000 tons, with the damage valued at over USD90 million.

The WEMA project, which ran from 2008 to 2018, addressed these constraints through integrated technology development and deployment of drought-tolerant and insect-resistant maize varieties in five African countries: Kenya, Mozambique, South Africa, Tanzania and Uganda. It was a public-private partnership between the Monsanto Company (subsequently acquired by Bayer in 2018) and the African

Agricultural Technology Foundation (AATF; see Box 4.1), the International Maize and Wheat Improvement Center and the national agriculture research systems in the respective countries.

The WEMA public-private partnership used a combination of classical breeding, marker-assisted breeding and transgenic technology to develop and deploy royalty-free drought-tolerant and insect-resistant ('climate-smart') white maize varieties to farmers in Africa. Under moderate drought conditions, the improved varieties are expected to increase yields by 20–35%.

> **Box 4.1 Catalysing and Driving Biotech Applications: The African Agricultural Technology Foundation**
> Established in 2003, AATF is a not-for-profit foundation, managed and led by Africans to foster public-private partnerships for the access and delivery of appropriate technologies for improved agricultural productivity. A unique role that AATF plays in enhancing access to technologies is through facilitating the acquisition of technologies from technology owners through royalty-free licenses or agreements, along with associated materials and knowhow for use on behalf of Africa's resource-poor farmers.
>
> AATF works on 6 commodities (maize, peanut, rice, banana, cowpea and cassava) in 13 countries, namely: Ethiopia, Kenya, Uganda and Tanzania in eastern Africa; Malawi, Mozambique, South Africa, Zambia and Zimbabwe in southern Africa; and Burkina Faso, Ghana, Nigeria and Senegal in West Africa.

The project developed several product options to suit different farmers' needs and requirements: conventional drought-tolerant (CDT) hybrids (marketed under the trade name DroughtTEGO®), CDT and insect-pest protected (*Bt*) hybrids and CDT, transgenic *Bt* and drought-tolerant stacked hybrids (marketed under the trade name TELA®). The project's reported impacts and achievements include as follows:

- Seventy-four conventionally bred DroughtTEGO® hybrids have been released across the 5 countries, and 46 hybrids have been commercialized in Kenya with a recorded average yield advantage of over 3 tons per hectare. Specifically, DroughtTEGO® varieties yielded an average of 4.9 tons/hectare under good agronomic practices as compared to 3.2 tons/hectare for commercial checks and 1.7 tons/hectare for the Kenyan national yield average (Muinga et al. 2019).
- Confined field trials for *Bt* maize hybrids have been carried out in Kenya, Uganda and South Africa.
- In 2015, general release of the drought-tolerant trait was approved in South Africa; planting stacked *Bt* and drought-tolerant hybrids was approved, and now five MON89034 *Bt* hybrids are available to seed companies in the country.

- In 2016, conditional general release and the conduct of national performance trials were approved in Kenya.
- In 2016, planting of stacked *Bt* and drought-tolerant hybrids in Kenya and Uganda was approved.

The combined Uganda and Kenya five season data for MON810 containing field trials reveal that the *Bt* gene had a significant effect on yields across varieties and trials with an average yield increase of 52%. A total of 34 hybrids evaluated for 3 years across 5 sites in 3 countries and incorporating 5 transgene hybrids gave 8–14% yield increase compared to varieties without these traits. An adoption study carried out in Kenya in 2017 revealed that within 3 years of dissemination of the DroughtTEGO® hybrids, 61% of the farmers surveyed were aware of and 26% had adopted the varieties.

Source: DeGroote (2011); Oikeh (2016)

4.5.5   Aflasafe™: Mitigating Aflatoxin Risks

Aflatoxin, a naturally occurring mycotoxin, is a poison produced by a soil fungus, primarily *Aspergillus flavus*. It infects crops, predominantly grains and legumes, threatening food security, health and trade in many developing countries, contaminating the food value chain and costing hundreds of millions of dollars each year in export opportunities. Aflatoxin contamination of several crops is common in tropical and subtropical regions. Contamination occurs across the production, postharvest and primary processing steps in the value chain.

Maize and groundnut, staples for hundreds of millions of people in Africa, are among the most susceptible to aflatoxin contamination. Susceptible crops apart from maize and groundnut include tree nuts, chilli peppers, sorghum, sesame seed and figs, among others. A stealthy and silent killer, aflatoxin is a major concern because of its acute, chronic and irreversible health effects on people and livestock, some-times leading to fatalities. Globally, an estimated 25% of aflatoxin-prone crops are contaminated with aflatoxins and/or other mycotoxins (Eskola et al. 2019).

In Africa specifically, aflatoxin is known to cause stunting in children and an estimated 30% of all liver cancer cases. Besides being life-threatening and compromising health, aflatoxin contamination hampers domestic, regional and international trade as companies are unable to meet international and regional standards. Farmers producing contaminated crops therefore cannot sell to premium markets, including export markets. Acute exposure results in mortality rates of 25–40%, while chronic exposure of lower levels causes impaired immune function, growth retardation and liver disease (Gong et al. 2002). Due to the effects of greater climate variability, areas that will see increases in rainfall and ambient moisture levels are expected to become increasingly more susceptible to aflatoxin contamination.

A collaborative initiative between the International Institute of Tropical Agriculture (IITA) and the US Department of Agriculture has developed several biocontrol products, marketed under the trade name Aflasafe™, to combat aflatoxin contamination in staple cereal crops and groundnuts. These products were developed by identifying 'friendly' fungi that are highly effective at reducing aflatoxin levels and then testing them in farmers' fields. This testing helps create the best composition of Aflasafe™ for each country while also providing data needed for the product's registration and regulatory approval. With Dalberg Global Development Advisors (supporting development of scaling strategies for specific countries), Chemonics—a private international development firm (operationalizing rollout strategies on the ground)—and IITA (providing the research and development knowledge) implemented the Aflasafe™ Technology Transfer and Commercialization (ATTC) project with funding from the Bill and Melinda Gates Foundation and USAID. ATTC commercialized and delivered Aflasafe™ to the marketplaces of Nigeria, Senegal, the Gambia, Burkina Faso, Ghana, Kenya and Tanzania. The project researched the potential for moving Aflasafe™ from a research to a commercial product, identified private sector investors, transferred the technology to selected investors and supported commercialization so that Aflasafe™ manufacturing and distribution systems could be economically viable, sustainable and independent. By collaborating with private sector and government stakeholders, ATTC worked to improve food safety and increase the income of maize and groundnut smallholder farmers through Aflasafe™ use.

The scaling process in each country is complex and time-consuming. It starts with the production of country-specific product. Each product contains four atoxigenic isolates native to the target country; in some cases, multiple products are produced for one country. The second step of the process is product registration. After intensive laboratory and field tests, dossiers for registration are prepared and submitted to biopesticide regulatory agencies. Aflasafe™ was first registered in Nigeria. In total, Aflasafe™ products have been registered with pesticide regulatory authorities for use in ten African nations: Nigeria (2014), Kenya (2015), Senegal and the Gambia (2016), Burkina Faso (2017), Ghana (2018), Zambia (2018), Tanzania (2018), Mozambique (2019) and Malawi (2020) (Konlambigue et al. 2020). In 2021 products are being developed for another ten countries. To overcome challenges for adoption of Aflasafe™, a project was implemented in Nigeria (2014–2019) that promoted large-scale adoption through an incentivization scheme in the country.

The third step of the scaling process involved licensing and distribution, while the fourth was the establishment of the manufacturing plant for mass production. By 2020, there were three operational Aflasafe™ manufacturing plants—Ibadan, Nigeria; Katumani, Kenya; and Kahone, Senegal—while construction of a plant in Arusha, Tanzania, was underway. The fifth step of the scaling process was commercialization, which began once registration was granted. ATTC has successfully transferred Aflasafe™ technology to manufacturers and distributors in Nigeria, Kenya, Senegal, the Gambia, Burkina Faso, Ghana and Tanzania, allowing local and regional exporters, buyers and smallholder farmers to access the product.

After more than 15 years of research, IITA, together with national institutions in these countries and international technical and development partners, has been able to successfully create a working biocontrol solution to address aflatoxin contamination for use in maize (ten countries), groundnut (nine countries) and sorghum (in Ghana). When properly applied, Aflasafe™ consistently reduces aflatoxins in these crops by 80–100%. The widespread application of Aflasafe™ in aflatoxin-affected areas can significantly increase quantities of aflatoxin-safe crops and reduce health effects, including liver cancer and potential stunting in children, while also increasing incomes of smallholder farmers.

4.5.6 Making Biofortification Work for Nutrition

Biofortification is a process by which crops are bred in a way that increases their nutritional value. Breeding nutritious crops is much cheaper than adding micronutrients to already processed foods. Lack of essential micronutrients such as iodine, iron, zinc and vitamin A in diets, often referred to as the hidden hunger—because the symptoms of deficiency often manifest only when they become severe—is a threat to millions of African lives.

Working with its partners, especially HarvestPlus, USAID has made significant contributions to the development and deployment of several biofortified commodities, which are making major contributions to human nutrition in and outside of Africa. Uganda (beans and sweet potato), Zambia (maize, sweet potato and cassava) and Rwanda (beans) are among countries that have benefited from the biofortification work and scaling of the resulting products.

The story of orange-fleshed sweet potato (OFSP) in Uganda (Box 4.2) is just one example of 13 currently available biofortified varieties of crops, including vitamin A-rich sweet potato, cassava and maize; iron-rich beans and pearl millet; and zinc-rich wheat and rice, which have been released in 30 countries around the world. Peer-reviewed clinical trials have demonstrated that biofortified foods have a positive impact on health and nutritional status, including reduction in the prevalence of diarrhoea among children under 3; reversal of iron deficiency in children and women; improved cognitive and physical performance in women and children; and improved night vision in children.

> **Box 4.2 USAID Support to Scaling of Biofortified Sweet Potato in Uganda**
>
> Vitamin A deficiency is among major health problems worldwide that leads to blindness, retarded growth and death, particularly in developing countries. In these countries, vitamin A deficiency largely affects preschool children, pregnant and lactating mothers and the rural poor. In SSA vitamin A deficiency affects 43 million children under age 5 and contributes to high rates of blindness, disease and premature death in children and pregnant women.

(continued)

Box 4.2 (continued)

Lack of vitamin A also impedes children's growth, increases their vulnerability to disease and contributes to poor immune function and maternal mortality.

The predicted prevalence of vitamin A deficiency for 36 SSA countries is 19.1% (Gurmu et al. 2014). In Uganda, it affects 38% of children aged 6–59 months and 36% of women aged 15–49 years, according to the 2011 Uganda Demographic and Health Survey.

Sweet potato is a low-priced crop which is a staple food in many SSA countries. Most varieties of sweet potato grown in Africa in the past had white, cream or yellow flesh and contained little or no provitamin A, but OFSP contain much higher quantities, principally of beta-carotene. The orange-fleshed variety of sweet potato was introduced in Uganda in 2007 by HarvestPlus project of the CGIAR Program on Agriculture for Nutrition and Health. By including OFSP in the diet, it is possible to increase the bioavailability of different micronutrients such as iron, zinc, calcium and magnesium, reduce vitamin A deficiency and hence reduce child mortality rates by 23–30%. Only 125 g of a fresh sweet potato root from most orange-fleshed varieties contains enough beta-carotene to provide the daily provitamin A needs of a preschool child.

In addition to addressing vitamin A deficiencies, the OFSP is as sweet as the indigenous white sweet potato and has high and fast-maturing yields, and there is high interest by local communities to grow OFSP for home consumption. Orange-fleshed sweet potato is growing in popularity and becoming an important strategy to improve vitamin A deficiency across Uganda. Through a USAID supported program, OFSP had been adopted by over 55,000 Ugandan farming households by 2018, with up to 237,000 households expected to be planting and eating the crop that year. A total of six varieties have been released and in use in the country.

Agricultural extension staff work closely with farmers' groups and other parties to ensure widespread OFSP availability and sustainability. Nutrition training is also offered to the farmers. Schools in the country that grow sweet potatoes in their gardens as a main crop are increasingly adopting the OFSP. With support from USAID, HarvestPlus also worked on linking OFSP farmers to markets.

4.5.7   Maize Lethal Necrosis Disease

Maize lethal necrosis (MLN) is a disease of maize caused by co-infection of maize with maize chlorotic mottle virus (MCMV) and one of several viruses from the *Potyviridae* family of viruses, such as sugarcane mosaic virus, maize dwarf mosaic virus, Johnsongrass mosaic virus or wheat streak mosaic virus. These viruses are transmitted mainly by insects including aphids and beetles. The disease was first identified in many East African countries between 2011 and 2014. According to CIMMYT, the disease caused estimated losses of up to 23% of the average annual production in the region.

Several collaborative projects have been involved in efforts to contain MLN in eastern Africa. One collaboration was funded by USAID through the MLN Diagnostics and Management Project (2015–2019) jointly supported by a CGIAR Research Programme led by CIMMYT. The collaborative project brought together national, regional and international partners and aimed to prevent the spread of MLN-causing MCMV, to support seed companies to produce MCMV-free commercial seed and to facilitate wider use of improved agronomic practices for disease control and create a phytosanitary community of practice for disease diagnosis and management. The project combined the use of tolerant varieties, crop rotation, introduction of maize-free seasons to reduce the pressure from the insect vector that transmits the viruses that cause MLN and the use of rapid MLN diagnostics. To implement MLN management approaches effectively, country-specific MLN management checklists were developed by over 160 seed industry players, including seed companies, breeders and regulators.

Major achievements have included development of 15 MLN-tolerant maize hybrids by 2015, 3 varieties of which had been released and 1 variety (Bazuka) commercialized in Uganda; an effective surveillance and monitoring system and emergency response plan for MLN surveillance developed, shared and put to use; enhanced technology diffusion achieved through sensitization and strengthening human and institutional capacities, leading to complete containment of MLN spread from affected countries; and an integrated pest management manual on MLN developed with a comprehensive protocol on how to produce MLN-free products. Due to the rapid detection of MLN and the response of the countries involved, it has been reported that the disease is under control. Farmers using MLN-tolerant varieties reportedly recorded additional yields of over 5 tons per hectare, worth an additional USD 1600 per hectare.

However, MLN has not been eradicated yet. Thus, it is important to continue monitoring to forestall possible re-emergence and spread within the region. Specific gaps that need to be addressed going forward include additional research on epidemiology of MLN to inform strategy development for control of infection—for example, there is a need to understand the development of MLN resistance in maize, which can then be exploited for screening; being a relatively new disease, there is a need to develop local capacity at institutional and individual levels for R&D and for strengthening sanitary and phytosanitary techniques and frameworks. Because MLN is a regional problem due to rapid spread of the disease across borders, it is important that research and technical guidance are coordinated through partnership at regional level. In eastern Africa, the epidemic is aggravated by very limited access to MLN-tolerant maize varieties by farmers and year-round cultivation of maize, allowing transmission via insect vectors.

4.5.8 Biotechnology Options for Managing Striga in Cereals in Africa: Lessons from the *Desmodium* Genus

Striga or witchweeds, comprising obligate root parasites in the genus *Striga*, are a major constraint to efficient production of cereal crops in Africa, including maize, sorghum, finger millet and upland rice. The main species attacking these crops in the continent are *Striga hermonthica* and *S. asiatica* (Orobanchaceae), with the former—infecting approximately 40% of arable land on the African savannah—being more economically important.

Striga is so adapted and integrated with its host plants that its seeds will only germinate from induction by specific chemical signals released by roots of these and some non-host plants. Once striga seed germinates, it extends its radicle towards the host roots and develops a special invasive organ, the haustorium, at the tip of the radical that the parasite uses to attach onto the roots of the host plants. Through this root-to-root attachment, striga inhibits normal host growth via three processes: absorbing its supply of moisture, nutrients and minerals; weakening of the host, wounding its outer root tissues and impairing photosynthesis; and a phytotoxic effect on the host within days of attachment (Gurney et al. 2006). This results in a large reduction in host plant growth and development and causes up to 100% yield loss in maize, severely affecting harvests of smallholder, resource-poor farmers. Striga thus directly affects livelihoods of approximately 300 million people in the continent.

The striga menace continues to increase in geographic coverage and in level of infestation. Striga control is, however, complicated by the abundant seed production by the parasite (up to 200,000 seeds per plant), longevity of the seed bank in the soil (up to 10 years) and the complicated mode of parasitism.

In the quest to develop effective management of lepidopteran stemborers in maize, a collaboration between the International Centre of Insect Physiology and Ecology (ICIPE), Rothamsted Research (UK) and Kenya Agricultural and Livestock Research Organisation (KALRO, formerly KARI) serendipitously discovered effective control of striga through intercropping with forage legumes in the genus *Desmodium* (Khan et al. 2002). Subsequent studies, funded by the Gatsby Charitable Foundation (UK), the Rockefeller Foundation, Kilimo Trust East Africa, European Union and UK Biotechnology and Biological Sciences Research Council, among others, revealed that *Desmodium* suppressed striga through a range of mechanisms, of which an allelopathic effect was the most important. A number of novel isoflavanones were elucidated from the active root exudates of *Desmodium*, some of which induced striga seed germination, while others inhibited radicle growth (Tsanuo et al. 2003; Hooper et al. 2015). *Desmodium* thus prevents parasitism of maize through a combination of stimulating germination and interfering with the subsequent development of striga. This results in suicidal germination of the parasite, as well as progressive depletion of the seedbank in the soil (Khan et al. 2008). This, together with *Desmodium*'s ability to improve soil health and effective control of stemborer pests, results in significant increases in grain yields (Midega et al. 2014). In a study, maize intercropped with *Desmodium* yielded about 3 tons per hectare, 140% higher than maize planted in sole plots (Midega et al. 2014).

The biosynthetic pathway of the C-glycosylflavones that are important in striga suppressions by *Desmodium*, including isoschaftoside (6-C-α-L-arabinopyranosyl-8-C-β-D-glucopyranosylapigeni), has been elucidated and shown to be initiated by a C-glycosyltransferase. This biosynthetic pathway is already mostly present in edible legumes and in cereals (Hooper et al. 2009, 2010). There is, therefore, an opportunity now to produce edible legumes, through feasible breeding programmes, suitable for intercropping with maize and other cereals, or cereals themselves by heterologous gene expression, to respond to a broader profile of farmer practices.

4.5.9   Exploiting Innate Plant Defence for Control of Maize Stemborers: Lessons from Grasses

Lepidopteran stemborers are major obstacles to efficient production of maize in Africa, where they cause grain yield losses of up to 88% (Kfir et al. 2002). One of the most injurious stemborer species is *Chilo partellus* (Swinhoe), an invasive moth species first detected in Africa sometime around 1930 that has since spread to many different agroecological zones in the continent. The pest uses components of the volatile organic compounds (VOCs) that plants emit to locate its hosts. A gravid female moth then lays eggs in batches on the plants that hatch into larvae, which are the damaging stage of the pests. The young larvae initially feed on leaves of the plants before boring into the stem. Effective control of these pests continues to elude most smallholder farmers who cannot afford the recommended chemical control strategies. Furthermore, because the larvae burrow into the plant stems, most control methods do not reach them. Moreover, effectiveness of many cultural methods is questionable, resulting in majority of smallholder farmers not deploying any control methods against these pests.

Maize, just like any other plant, has evolved sophisticated defence strategies against attack by pests; indirect defence that involves attraction of the pests' natural enemies is one of the most important. These natural enemies have similarly evolved chemosensory systems that are fine-tuned for recognizing plants on which their prey is present (Bruce et al. 2010). Insect attack induces secondary metabolism in plants, leading to increased emission of volatiles, known as herbivore-induced plant volatiles (HIPVs), that serve as a 'cry for help' with two consequences: they attract the pests' natural enemies (de Moraes et al. 1998) and reduce further colonization by the pests (Kessler and Baldwin 2001). Such tritrophic interactions involving plants, insect pests and the natural enemies are nature's way of balancing coexistence of biological diversity.

In the quest to find more effective ways to manage stemborer pests, scientists at ICIPE in Kenya in collaboration with colleagues at Rothamsted Research (UK) and other partners discovered a 'smart' trait in a forage grass that involved changes in plant chemistry induced by the earliest stage of pest attack, oviposition. They discovered that oviposition by *C. partellus* on signal grass, *Brachiaria brizantha*, led to a marked decrease in emission of VOCs utilized by stemborer moths to locate host plants, notably (Z)-3-hexenyl acetate, and increased ratio of other volatiles.

Two observations of ecological importance were made: firstly, the new blend reduced further colonization by the pests; and secondly, this led to increased attractancy of the plants to the parasitoid *Cotesia sesamiae*, a natural enemy that attacks stemborer larvae. The novelty of the discovery were that a plant was responding to a non-damaging stage of the pest through suppressed emission of pest-attracting volatiles that altered the compound ratios in the resultant blend; a specialist natural enemy, through 'eavesdropping', utilized the new blend to locate a plant on which its prey was present; and a plant bearing eggs attracted parasitoids of the next (damaging) stage of pest attack, larvae, before actual damage occurred on the plant (Bruce et al. 2010). This 'early herbivore alert' signalled a finely tuned and co-evolved defence response sensitive even to the earliest stage of pest attack (Hilker and Meiners 2006).

This 'smart' trait was subsequently discovered in teosinte, the ancestor of maize (Mutyambai et al. 2015), and also maize landraces from South America and open pollinated varieties used in East Africa (Tamiru et al. 2011, 2012, 2015, 2020). Notably, while there was suppressed emission of VOCs with *B. brizantha*, there was increased emission of HIPVs in the teosinte and maize germplasm. These plants attracted both the egg and larval parasitoids in addition to reducing further attack by the pest. However, the 'smart' trait was rare among the elite maize hybrids, implying these naturally occurring defence responses might have been lost during selective breeding for traits such as grain yield in majority of the hybrids. More recently, investigations into the genetic variability in induced responses to stemborer oviposition in maize through a genome-wide association study have revealed maize genomic regions that are associated with the 'smart' indirect defence (Tamiru et al. 2020). Additionally, marker-trait associations have been identified, some of which are linked to genes involved in plant defence. This now paves the way for selection of this trait that exploits natural plant defence by maize breeders, together with an opportunity for its introgression into elite hybrid varieties with higher yield potential. This would provide a cost-effective and environmentally benign control option, and with better maize resistance against stemborer attack, for smallholder farmers in Africa and beyond.

References

Alhassan WS (2001) The status of agricultural biotechnology in selected West and Central African countries. International Institute of Tropical Agriculture, Ibadan

Bailey R, Willoughby R, Grzywacz D (2014) On trial: agricultural biotechnology in Africa. Energy, environment and resources, Chatham House

Bandyopadhyay R, Ortega-Beltran A, Akande A et al (2016) Biological control of aflatoxins in Africa: current status and potential challenges in the face of climate change. World Mycotoxin J 9(5):771–789. https://doi.org/10.3920/WMJ2016.2130

Bavier J (2017) How Monsanto's GM cotton sowed trouble in Africa. Reuters Investigates. December 8 2017. Available via Reuters: https://www.reuters.com/investigates/special-report/monsanto-burkina-cotton/. Accessed 30 June 2021

Bouis H, Saltzman A, Low J, Ball A, Covic N (2017) An overview of the landscape and approach for biofortification in Africa. Afr J Food Agric Nutr Dev 17(2):11848–11864. https://doi.org/10.18697/ajfand.78.HarvestPlus01

Brink JA, Woodward BR, DaSilva EJ (1998) Plant biotechnology: a tool for development in Africa. Electron J Biotechnol 1(3):142–151

Bruce TJA, Midega CAO, Birkett MA, Pickett JA, Khan ZR (2010) Is quality more important than quantity? Insect behavioural responses to changes in a volatile blend after stemborer oviposition on an African grass. Biol Lett 6:314–317

Chambers JA et al (2014) GM agricultural technologies for Africa: a state of affairs. Report of a study commissioned by the African Development Bank, IFPRI and AfDB, Washington and Abidjan

De Moraes CM, Lewis WJ, Pare PW, Alborn HT, Tumlinson JH (1998) Herbivore-infested plants selectively attract parasitoids. Nature 393:570–573

Dowd-Uribe B, Schnurr MA (2016) Briefing: Burkina Faso's reversal on genetically modified cotton and the implications for Africa. Afr Aff 115(458):161–172

Edwards JD, McCouch SR (2007) Molecular markers for use in plant molecular breeding and germplasm evaluation. In: FAO, Guimaraes EP, Ruane J et al (eds) Marker-assisted selection: Current status and future perspectives in crops, livestock, forestry and fish. FAO, Rome

Eskola M, Kos G, Elliott CT et al (2019) Worldwide contamination of food-crops with mycotoxins: validity of the widely cited 'FAO estimate' of 25%. Crit Rev Food Sci Nutr 30:1–17

FAO (2021) Global partnership initiative for plant breeding capacity building. FAO, Rome. Available via FAO: www.fao.org/in-action/plant-breeding/background/zh/. Accessed 30 June 2021

FARA (Forum for Agricultural Research in Africa) (2011) Status of biotechnology and biosafety in sub-Saharan Africa: a FARA 2009 study report. FARA Secretariat, Accra

Farrow A, Muthoni-Andriatsitohaina R (eds) (2020) Atlas of common bean production in Africa, 2nd edn. PABRA/CIAT, Nairobi, p 260. ISBN: 978958694227-0

Goedde L, Ooko-Ombaka A, Pais G (2019) Winning in Africa's agricultural market. Our insights. McKinsey and Company Available via McKinsey: https://www.mckinsey.com/industries/agriculture/our-insights/winning-in-africas-agricultural-market. Accessed 26 Oct 2021

Gong YY, Cardwell K, Hounsa A et al (2002) Dietary aflatoxin exposure and impaired growth in young children from Benin and Togo: cross sectional study. Br Med J 325:20–22. https://doi.org/10.1136/bmj.325.7354.20

Gurmu F, Shimelis H, Laing MD (2014) The potential of orange-fleshed sweet potato to prevent vitamin a deficiency in Africa. Int J Vitam Nutr 84(1–2):65–78

Gurney AL, Slate J, Press MC et al (2006) A novel form of resistance in rice to the angiosperm parasite Striga hermonthica. New Phytol 169:199–208

Habte E, Katung E, Yirga C et al (2021) Adoption of Common bean technologies and its impacts on productivity and household welfare in Ethiopia: lessons from tropical legumes project. Ethiopian Institute of Agricultural Research, Addis Ababa. https://doi.org/10.13140/RG.2.2.17130.03524

Hilker M, Meiners T (2006) Early herbivore alert: Insect eggs induce plant defense. J Chem Ecol 32:1379–1397

Hooper AM, Hassanali A, Chamberlain K et al (2009) New genetic opportunities from legume intercrops for controlling Striga spp. parasitic weeds. Pest Manag Sci 65:546–552

Hooper AM, Tsanuo MK, Chamberlain K et al (2010) Isoschaftoside, a C-glycosylflavonoid from desmodium uncinatum root exudate, is an allelochemical against the development of striga. Phytochemistry 71:904–908

Hooper AM, Caulfield JC, Hao B et al (2015) Isolation and identification of Desmodium root exudates from drought tolerant species used as intercrops against *Striga hermonthica*. Phytochemistry 117:380–387

IFPRI (2014) Africa's next harvest II: biotechnology R&D driving agriculture productivity and economic development (Kenya). International Food Policy Research Institute, Washington, DC

ISAAA (2015a) Biotech country facts and trends: Burkina Faso. ISAAA SEAsia Centre, Phillipines

ISAAA (2015b) Biotech country facts & trends: South Africa. ISAAA SEAsia Centre, Phillipines

ISAAA (2016) Global Status of Commercialized Biotech/GM Crops: 2016. ISAAA Brief No. 52. ISAAA, Ithaca

ISAAA (2019) Global Status of Commercialized Biotech/GM Crops in 2019. ISAAA Brief No. 55. ISAAA, Ithaca

Kabunga NS, Dubois T, Qaim M (2012) Yield effects of tissue culture bananas in Kenya: accounting for selection bias and the role of complementary inputs. J Agric Econ 63:444–464

Katungi E, Nduwarigira E, Ntukamazina N et al (2020) Food security and common bean productivity: impacts of improved bean technology adoption among smallholder farmers in Burundi. The Alliance of Bioversity International and CIAT, Nairobi. Available via CGSPACE: https://cgspace.cgiar.org/handle/10568/109119. Accessed 26 Oct 2021

Kessler A, Baldwin IT (2001) Defensive function of herbivore-induced plant volatile emissions in nature. Science 291:2141–2144

Kfir R, Overholt WA, Khan ZR et al (2002) Biology and management of economically important lepidopteran cereal stem borers in Africa. Annu Rev Entomol 47:701–731

Khan ZR, Hassanali A, Overholt W et al (2002) Control of witchweed Striga hermonthica by intercropping with Desmodium spp., and the mechanism defined as allelopathic. J Chem Ecol 28:1871–1885

Khan ZR, Pickett JA, Hassanali A et al (2008) Desmodium species and associated biochemical traits for controlling Striga species: present and future prospects. Weed Res 48:302–306

Kikulwe EM, Kabunga NS, Qaim M (2012) Impact of tissue culture banana technology in Kenya: a difference-in-difference estimation approach. Discussion Papers, No. 117, Georg-August-Universität Göttingen, Courant Research Centre – Poverty, Equity and Growth (CRC-PEG), Göttingen

Kiome R (2015) A strategic framework for transgenic research and product development in Africa: Report of a CGIAR Study. ILRI, Nairobi

Konlambigue M, Ortega-Beltran A, Bandyopadhyay R et al (2020) Lessons learned on scaling Aflasafe® through commercialization in sub-Saharan Africa. International Food Policy Research Institute, Washington DC. Available via IFPRI: https://www.ifpri.org/publication/lessons-learned-scaling-aflasafe%C2%AEthrough-commercialization-sub-saharan-africa. Accessed 27 June 2021

Kyei F, Puobi R, EGadu1 S et al (2017) The science, acceptance and support of modern biotechnology in Africa. J Adv Biol Biotechnol 12(2):1–14

Masiga CW, Mneney E, Wachira F et al (2013) Situational analysis of the current state of tissue culture application in the Eastern and Central Africa region. Association for Strengthening Agricultural Research in East and Central Africa (ASARECA), Entebbe Available via ASARECA: https://wwwasarecaorg/publication/situation-analysis-current-status-tissue-culture-application-eastern-and-central-africa. Accessed 29 Oct 2021

Midega CAO, Salifu D, Bruce TJ et al (2014) Cumulative effects and economic benefits of intercropping maize with food legumes on Striga hermonthica infestation. Field Crops Res 155:144–152

Midling MB (2011) Biotechnology adoption in Africa. Berkeley Undergraduate J 24(3):93–111

Moyo M, Bairu MW, Amoo SO et al (2011) Plant biotechnology in South Africa: micropropagation research endeavours, prospects and challenges. S Afr J Bot 77(4):996–1011

Mtui GY (2011) Status of Biotechnology in Eastern and Central Africa. Biotechnol Mol Biol Rev 6(9):183–198

Muinga G, Marechera G, Macharia I et al (2019) Adoption of climate-smart DroughtTEGO® varieties in Kenya. Afr J Food Agric Nutr and Dev 19:15090–15108

Mulder M (2003) National biotechnology survey. eGoliBio Life Sciences Incubator, Pretoria

Mulugo L, Kyazze FB, Kibwika P et al (2020) Seed security factors driving farmer decisions on uptake of tissue culture banana seed in Central Uganda. Sustainability 12(23):10223. https://doi.org/10.3390/su122310223

Mutyambai DM, Bruce TJA, Midega CAO et al (2015) Responses of parasitoids to volatiles induced by Chilo partellus oviposition on teosinte, a wild ancestor of maize. J Chem Ecol 41: 323–329

Muyanga M (2009) Smallholder adoption and economic impacts of tissue culture banana in Kenya. Afr J Biotechnol 8(23):6548–6555

Njuguna M, Wambugu F, Acharya S et al (2010) Socio-economic impact of tissue culture banana (Musa Spp.) in Kenya through the whole value chain approach. In: Dubois T, Hauser C, Staver D et al (eds) Proceedings of the international conference on banana and plantain in Africa: Harnessing international partnerships to increase research impact, 5–9 October, 2008, Mombasa, Kenya. Acta Hortic, pp 77–86

Olembo N, M'mboyi F, Oyugi K et al (2010) Status of crop biotechnology in sub-Saharan Africa. African Biotechnology Stakeholders Forum, Nairobi

Oratungye KE, Rubyogo JC, Onyango P (2021a) Improving bean production and marketing in Africa (IBPMA) April 2020 – March 2021 Report, p 25. Alliance Bioversity CIAT, Rome. Available at https://alliancebioversityciat.org/publications-data/improving-bean-production-and-marketing-africa-ibpma-april-2020-march-2021-report. Accessed 28 Oct 2021

Oratungye, KE, Rubyogo JC, Onyango P (2021b) Improving food security, nutrition, incomes, natural resource base and gender equity for better livelihoods of smallholder households in sub-Saharan Africa, p 80. Available via CGSPACE: https://cgspace.cgiar.org/handle/10568/113988. Accessed 28 Oct 2021

Ouma E, Dubois T, Kabunga N et al (2013) Adoption and impact of tissue culture bananas in Burundi: an application of a propensity score matching approach. In: Blomme G, van Asten P, Vanlauwe B (eds) Banana systems in the humid highlands of sub-Saharan Africa. CABI, Oxfordshire, pp 2016–2223

Qaim M (1999) Assessing the impact of banana biotechnology in Kenya. ISAAA Briefs 10. ISAAA, Ithaca, New York

Quain MD, Asibuo JY (2009) Biotechnology for agriculture enhancement in Ghana: the challenges and opportunities. Asian Biotechnol Dev Rev 11(3):49–61

Showalter AM, Heuberger S, Tabashnik BE et al (2009) A primer for using transgenic insecticidal cotton in developing countries. J Insect Sci 9(1):22. https://doi.org/10.1673/031.009.2201

Tamiru A, Bruce TJ, Woodcock CM et al (2011) Maize landraces recruit egg and larval parasitoids in response to egg deposition by a herbivore. Ecol Lett 14:1075–1083

Tamiru A, Bruce TJ, Midega CA et al (2012) Oviposition induced volatile emissions from African smallholder farmers' maize varieties. J Chem Ecol 38:231–234

Tamiru A, Khan ZR, Bruce TJA (2015) New directions for improving crop resistance to insects by breeding for egg induced defence. Curr Opin Insect Sci 9:51–55

Tamiru A, Paliwal R, Manthi SJ et al (2020) Genome wide association analysis of a stemborer egg induced "call-for-help" defence trait in maize. Sci Rep 10:11205

Traore VSE (2016) BT cotton in Burkina Faso: the impacts and lessons. Was GMOs a necessary intervention? National Conference on "Just Governance: The Nigerian Biosafety Act, GMOs and Implications for Nigerians and Africa", Abuja May 24–25, 2016. Available via AFJN: http://aefjn.org/wp-content/uploads/2016/05/Edgar_Present_Conference_Abuja.pdf

Tsanuo MK, Hassanali A, Hooper AM et al (2003) Isoflavanones from the allelopathic aqueous root exudates of Desmodium uncinatum. Phytochemistry 64:265–273

UNDP, UNESCO (1994) African Network of Microbiological Resources Centres (MIRCENs) in biofertilizer production and use. Project findings and recommendations, UNDP, Paris

USDA (2016) Biotechnology frequently asked questions (FAQs). In U.S. Department of Agriculture. Available via USDA: https://www.usda.gov/topics/biotechnology/biotechnology-frequently-asked-questions-faqs. Accessed 29 June 2021

Virgin I, Bhagavan M, Komen J et al (2007) Agricultural biotechnology and small-scale farmers in Eastern and Southern Africa. Stockholm Environment Institute, Stockholm

Walker T, Alene A, Ndjeunga J et al (2014) Measuring the effectiveness of crop improvement research in sub-Saharan Africa from the perspectives of varietal output, adoption, and change:

20 crops, 30 countries, and 1150 cultivars in farmers' fields. Report of the Standing Panel on Impact Assessment (SPIA), CGIAR Independent Science and Partnership Council (ISPC) Secretariat, Rome

The State of Applications and Impacts of Biotechnology in the Livestock Sector

5

J. E. O. Rege, Joel W. Ochieng, and Keith Sones

Abstract

Application of biotechnology in the livestock sector in sub-Saharan African countries was considered based on both the extent (e.g. limited use in laboratories versus widespread field use) and context of use (e.g. cloning individual genes versus cloning a whole organism). The biotechnologies were classified in three groups, namely, low-, medium- and high-tech, depending on their relative complexity. As high-tech applications are only used in a few countries, low- and medium-tech were the most important for categorizing most countries. No countries were categorized as having 'very high' use of biotechnology applications in livestock, and just two, Kenya and South Africa, were ranked as 'high'. A further ten countries fell into the 'medium' use category, namely, Côte d'Ivoire, Ethiopia, Ghana, Mali, Mauritius, Namibia, Nigeria, Sudan, Tanzania and Uganda. The remaining 33 countries for which information was available were categorized as 'low' or 'very low'. Overall, livestock comes second to crops in terms of level of biotechnology applications in SSA.

J. E. O. Rege (✉)
Emerge Centre for Innovations-Africa, Nairobi, Kenya
e-mail: ed.rege@emerge-africa.org

J. W. Ochieng
Agricultural Biotechnology Programme, University of Nairobi, Nairobi, Kenya

K. Sones
Keith Sones Associates, Banbury, UK

5.1   Livestock as an Important Component of Agriculture in Sub-Saharan Africa

Africa is a livestock-rich continent representing about one-third of the world's livestock population, consisting (as at 2018) of 2 billion birds (1.9 billion chickens, 26 million guinea fowl, 27 million turkeys, 22 million ducks and 11.5 million pigeons), 438 million goats, 384 million sheep, just under 356 million cattle, 40.5 million pigs, almost 31 million camels and 38 million equines (including 30 million donkeys, 6.5 million horses and 885,000 mules). The livestock sector contributes between 30 and 80% of agricultural GDP of individual SSA countries (Malabo Montpellier Panel 2020).

Across the continent, livestock is considered as one of the most valuable agricultural assets for the rural and urban poor, especially for women and pastoralists. Across all SSA countries, poultry, particularly chicken, farming is a major agricultural activity, especially for women. Like in most developing regions of the world, livestock owners in SSA are predominantly resource-poor, small-scale operators with little or no land and few animals, who must operate within the constraints of the local climate and who have limited opportunity to determine resource allocation for animal production.

While poverty remains widespread throughout SSA, the continent is fast growing and the consumption of animal protein is growing fast, in particular for relatively low-value, low-processed livestock products. Overall, a small proportion of livestock keepers (between 5 and 20%, depending on the country) can be considered business-oriented with incentives to expand their livestock production and tap into the growing market for animal protein (World Bank 2014). They keep relatively large herds and derive a significant share of their cash income from sales of livestock or livestock products. The remainder of livestock keepers can be defined as 'livelihood-oriented': they keep animals more for the many livelihoods services they provide—such as insurance, manure and hauling services—than for selling meat, milk and other livestock products to the market (World Bank 2014). Policies and investments aimed at enhancing the contribution of livestock to economic growth and poverty reduction should therefore adopt a dual strategy of targeting livelihood-oriented and business-oriented livestock keepers, who have diverse incentives to keep animals.

It is widely recognized that growth of agriculture and agribusiness, including the livestock industry, is key to reducing poverty. As a large share of Africa's poor is made up of smallholder farmers, the majority of whom keep animals, increased livestock productivity can improve the livelihoods of these producers. Higher production also translates into lower prices of livestock products, to the benefit of the majority of households—rural and urban—who are buyers of animal protein. Meat and milk are sources of high-quality protein and essential amino acids, minerals, fats and fatty acids, readily available vitamins, small quantities of carbohydrates and other bioactive components.

As is the case in many parts of the world, meat is among the least affordable food options in SSA. It is generally more costly than locally available grains, beans, vegetables and fruit. However, as average incomes rise, more people start to eat increasingly more meat. Like in many developing regions, per capita consumption of animal-sourced foods in Africa is still way below the recommended levels to support good health. The continent is set to experience an unprecedented increase in demand for animal-sourced foods over the coming decades: demand for meat, milk and eggs in Africa will almost quadruple by 2050, fuelled by a ballooning population—expected to double to 2.4 billion people—and a growing appetite for high-protein foods driven by rising living standards.

Deployment of available and emerging biotechnologies can improve livestock productivity through interventions targeting enhanced reproduction, genetic improvement, animal health and welfare and improved nutritional quality and safety of animal-derived foods, among others.

5.2   Relative Levels of Applications of Biotechnology in Livestock

Advances in science, especially in the last few decades, have produced arrays of biotechnologies with relevance to livestock production in four broad areas: animal reproduction, genetics and breeding, animal nutrition and animal health. Specific emerging biotechnology opportunities for livestock include molecular cloning of genes, gene transfer, genetic manipulation of animal embryos, genetic manipulation of rumen microbes, chemical and biological treatment of low-quality animal feeds for improved nutritive value, genetically engineered immunodiagnostic and immunoprophylactic agents and recombinant veterinary vaccines.

This section reviews the status of SSA countries in regard to application of the range of biotechnologies relevant for livestock, from low-tech, through medium-tech to high-tech (Table 5.1). The descriptions of individual technologies in each category examines status of development and application—from those still restricted to research applications through those in limited experimental use to those widely used either in large-scale commercial operations or in smallholder livestock systems.

Table 5.1 Livestock biotechnologies on the basis of which countries have been classified

Level of technology	Low-tech	Medium-tech	High-tech
Examples of technologies	• Conventional selective breeding and crossbreeding (on phenotypes) • Artificial insemination • Oestrous synchronization • Probiotics and prebiotics in feeds • Use of silage additives	• Sperm sexing • Embryo transfer • Embryo sexing • Use of genetic markers (e.g. in breeding—MAS, characterization), including PCR • ELISA and PCR diagnostic tests • Recombinant vaccines	• Cloning • Genetic modification/engineering

Examples from selected countries are given to indicate the level and extent (or spread and depth) of research and use where information was available.

5.2.1 Low-Tech Applications

Classical Within-Breed Selection Classical artificial selection is an established biotechnology where animal populations are subjected to iterations of evaluations for traits of economic and biologic value followed by mating among selected animals to generate individuals with improved performance in subsequent successive generations. While crop breeding efforts have made significant impact in SSA (see Chap. 4), the same cannot be said of livestock breeding. This is in significant part because animal recording, a prerequisite for the establishment of a systematic breeding programme, is essentially absent in SSA—save for the commercial livestock sector in South Africa and, to an even smaller scale, Kenya—where organized breed societies undertake systematic recording.

Like in South Africa, the history of milk recording in East Africa (more specifically Kenya) is closely linked to the establishment of breed societies. The first official milk recording scheme was started voluntarily by commercial dairy farmers in East Africa in 1949 (Mosi 1984). This scheme—known as the East Africa Milk Recording Service—covered Uganda and Tanzania as well but was more active in Kenya. Its operations were confined to large-scale farms. Despite this history, milk recording has remained confined in scope in Kenya and is nearly non-existent in Tanzania and Uganda. Despite subsequent failed institutional restructuring, leading to the establishments of the Kenya Livestock Breeders Organization (Kosgey et al. 2011) with a mandate to address identified challenges in the country, the scaling of recording has remained problematic, and this has been attributable to resource constraints but is mostly a reflection of inability of stakeholders to appreciate the benefits they can derive from country-driven selective breeding—compared to importation of 'superior genetics' from developed countries. Pilot recording schemes have also been tried in a number of other countries, but these too have not led to the establishment of sustained recording systems.

In traditional livestock systems in SSA, the process of evaluation is often based on unwritten assessments not accompanied by systematic recording. This, and associated inaccuracies of selection, is what is in widespread use in farmer-led selection across SSA. Consequently, progress attributable to within-breed selective breeding at national levels is low at best. In any case, the absence of recording also means tracking of progress attributable to selection is impossible.

The absence of recording in developing countries has been a subject of discussion for decades, and publications abound on the subject. In 2016, FAO prepared guidelines for development of integrated multipurpose animal recording systems with the objective of helping countries to design and implement such systems and to maximize the chances that they will be sustained (FAO 2016). Unless recording is owned and run by farmers themselves, it is unlikely that sustainable recording programmes can be achieved. This explains why the limited schemes remain restricted to the commercial farming sub-sector where owners recognize the value of, and are willing to pay for, recording. Understandably, almost all these systems are based on exotic breeds. Consequently, indigenous livestock owned and farmed by smallholders, which have high adaptability but generally low levels of productivity and could benefit from sustained genetic improvement, have not been the focus.

In an attempt to address this challenge, the International Livestock Research Institute (ILRI), through the African Dairy Genetics Gains (ADGG) programme, is supporting the establishment of farmer-driven on-farm pedigree and performance recording in eastern Africa (initially in Tanzania and Ethiopia but has been expanded to Uganda, Rwanda and Kenya). The performance data collected from ADGG has enabled genomic prediction and selection of top young bulls for breeding in Ethiopia and Tanzania. Unfortunately, this and similar programmes focus on crossbred cattle and not on indigenous breeds. Programmes and projects focusing on sustained genetic improvements among indigenous breeds have demonstrated that this approach can, indeed, be used to produce populations with superior genetic merit. Examples include the Improved Boran cattle and Red Maasai sheep in Kenya and the Horro chicken breeding programme as well the community sheep and goat breeding programme, both in Ethiopia (see Sects. 5.4.3 and 5.4.4).

The International Committee for Animal Recording (ICAR) aims to promote the development and improvement of performance recording and evaluation of livestock globally. African members of ICAR are Egypt, Tunisia and Morocco; no SSA country is currently a member, but the South African Stud Book and the Agricultural Research Council (ARC) of South Africa, while not members, have received ICAR's Certificate of Quality. ICAR is making efforts to engage African countries through training and workshop events. For example, in 2015, ICAR, in collaboration with FAO and other stakeholders, organized a symposium in Pretoria, 'African Symposium on animal identification and recording systems for traceability and livestock development in sub-Saharan Africa', at which 30 attending SSA countries made a declaration (the Pretoria Declaration) committing to setting up national animal recording and identification systems.

A promising approach to expanding animal recording in Africa is to use mobile phones and other rapidly developing information and communication technologies. Proof-of-concept of such approaches has been achieved for dairy cattle and small ruminants, although reliable Internet connectivity remains a challenge to their wide-scale use. A newly formed network of African animal breeding practitioners in Africa and in the diaspora—African Animal Breeding Network (AABNet)—has identified multi-country genetic gains as one of its pillars and plans to apply emerging tools such as phenomics in pursuit of this endeavour (Mrode et al. 2020). This represents the first time such a commitment has been made by an institution whose membership suggests that it has the collective skills and clout to be able to mobilize the goodwill and resources, including intellectual capital, to design and operationalize—even at a demonstration scale—what a functional breeding programme in the African production systems context might look like.

Crossbreeding and Formation of Composite Breeds Inadequate nutrition and presence of significant biotic and abiotic stresses, including poor management, continue to favour the use of indigenous livestock breeds in low-input smallholder systems in SSA. To date, however, there are hardly any success stories of within-breed improvements of indigenous breeds. Crossbreeding of indigenous livestock with exotic breeds has been seen as a faster way of making genetic improvements and one that does not require establishment of performance and pedigree recording. Relatively modest increases in output can lead to large gains in the efficiency with which feed resources are used, and this underlies the persistent attempts across many SSA countries to upgrade the indigenous breeds by using imported breeds with higher additive genetic performance, particularly for milk and meat production. Indeed, because of its relative technical simplicity, and faster although not necessarily sustainable results, it is more widespread than systematic classical selective breeding within indigenous livestock populations. It is applied in all major livestock species—cattle, sheep, goats, pigs and chickens. East Africa (especially Kenya, but also Tanzania, Uganda, Rwanda and Ethiopia to smaller extents) is ahead of other SSA regions in the use of crossbreeding for generating grade dairy cows used predominantly in smallholder systems. But crossbreeding is also widely applied in beef systems in southern Africa and sheep and goats as well as in chickens across the African continent.

In addition to the use of crossbreeding to generate high-grade livestock using different systems, such as single cross, backcross and rotational crossing, there has been increased interest in the creation of composite or synthetic breeds involving two or more parental breeds and apply a combination of planned crossbreeding and selection. Notable examples of composite breeds that have been developed from planned selective crossbreeding include the following: in South Africa, the Boer goat and several unique cattle breeds, such as the Sanganer (an Afrikaner-Nguni composite) and Bonsmara (a 5/8 Afrikaner and 3/8 Shorthorn or Hereford composite); the Mpwapwa dual purpose cattle of Tanzania; the Kenya dual purpose Goat; and the Shika brown and Funaab-alpha chicken breeds in Nigeria. Africa is also home to a large number of composite sheep breeds (Rasali et al. 2006), including the Mossi of Burkina Faso; Maroua of Cameroon; Nungua black head of Ghana; Permer and Yankassa of Nigeria; Worale of Senegal; Afrino, Dormer, Dorper, Meatmaster and Van Rooy of South Africa; Ingessana, Meidob and Toposa of Sudan; and Vogan of Togo.

Overall, crossbreeding, followed by conventional selection within the crossbred populations, is the most widely used breeding technologies across the continent with almost all countries applying these, albeit to varying extents in different livestock species, with cattle topping the list.

Breed Characterization Phenotypic characterization of indigenous breeds and breed comparisons involving both pure and crossbreds have been among the most popular subjects of study by animal breeders in Africa for decades. Initially predominantly focused on such production traits as growth, milk yield, carcass characteristics, egg production and reproductive traits, a shift in focus towards adaptation traits—disease, pest and heat resistance/tolerance—occurred intensely in the mid-1970s through the 1990s, but the arrival of molecular tools shifted emphasis away from phenotypic characterization.

Two thrusts stand out in SSA with regard to breed characterization. One was around trypanotolerance, led by the International Livestock Centre for Africa (ILCA) which was headquartered in Ethiopia, the International Laboratory for Research on Animal Diseases (ILRAD) headquartered in Kenya (the two centres merged in 1995 to form ILRI) and the International Trypanotolerance Centre based in The Gambia. Much of the effort was on the characterization of the N'Dama cattle and other humpless African taurine relatives that are resident in the humid parts of West and Central Africa (Rowlands and Teale 1994). Many projects were initiated to compare N'Dama and other indigenous breeds—including in eastern Africa. One such project led to the finding that the Orma Boran in Kenya (a zebu) and the Sheko (a humpless shorthorn) of south-west Ethiopia were more tolerant to trypanosomosis than the zebu breeds in these areas. These early results triggered a hunt for the genetic regions of the N'Dama responsible for the trypanotolerance trait (e.g. Hanotte et al. 2003).

The second initiative focused on genetic resistance to endoparasites. This was triggered by the finding by ILRI researchers (Baker et al. 1992, 2003) that the Red Maasai sheep of Kenya had innate resistance to gastrointestinal parasites. Again, this triggered a number of new breed comparison trials focusing on parasite resistance as well as a search for quantitative trait loci (QTLs) associated with endoparasites (Baker et al. 1998; Rege et al. 2002; Matika et al. 2003; Silva et al. 2012) and tick resistance (Mwangi et al. 1998; Wambura et al. 1998).

Interests in the N'Dama cattle and Red Maasai sheep, and their disease resistance attributes, are bound to increase as advances in molecular marker technologies and new analytic tools open new opportunities for deepening the understanding of the underlying genetic mechanisms of these attributes.

Reproductive Technologies Reproductive technologies covered in this section include artificial insemination (AI), oestrus synchronization, embryo transfer (ET), multiple ovulation embryo transfer (MOET), semen sexing, in vitro fertilization (IVF) and cloning. Table 5.2 summarizes the status of research in these technologies in Africa as at 2014 and captured in the Second State of the World Report on Animal Genetic Resources (FAO 2015), which is the most recent edition of this series.

Out of the 40 African countries which provided information (including North Africa), 43% and 30% indicated that they had some research on AI by national and international programmes, respectively. Corresponding figures for ET were 30 and 23%, while those for semen sexing (8 and 0%), in vitro fertilization (8 and 3%) and cloning (3 and 0%) were much lower. Indeed, there was only one case of cloning reported—in South Africa: the successful cloning by ILRI in Kenya described in Sect. 5.2.3 was not covered in the report. Out of the eight eastern African countries which reported, 50% and 63% had national research activities on AI and ET, respectively. In southern Africa, the figure was 25% for both AI and ET. Only Kenya and South Africa reported any research in semen sexing and in vitro

Table 5.2 Status of research in the use of reproductive technologies[a]

Region	No. of responding countries	% of responding countries reporting research on:									
		Artificial insemination		Embryo transfer or MOET		Semen sexing		In vitro fertilization		Cloning	
		Na.	Int.	Na.	Int.	Na.	Int.	Na.	Int.	Na.	Int.
Eastern Africa	8	50	25	63	50	13	0	13	13	0	0
Southern Africa	12	25	25	25	17	8	0	8	0	8	0
West and North Africa	20	50	35	20	15	5	0	5	0	0	0
Africa	40	43	30	30	23	8	0	8	3	3	0
World including Africa	128	64	35	49	30	28	15	37	20	18	12

Source: FAO (2015)

[a]Note: Na. = national institutions; Int. = international institutions

fertilization. Ten out of the 20 reporting countries in North and West Africa group reported research on AI, but only 1 country in the group reported some research on semen sexing and in vitro fertilization.

Artificial Insemination Among reproductive technologies, AI has perhaps been the most widely applied animal biotechnology in SSA and globally, especially where use of frozen semen has been possible. The major advantage of AI comes from the significant genetic improvement it can drive in productivity, as well as the rapid dissemination of sire-side genetic merit it can facilitate internationally. This makes it possible for superior genes to be used widely and intensively. Complementary technologies, such as the monitoring of reproductive hormones, heat detection, oestrus synchronization and semen sexing, can improve the efficiency of AI significantly.

Availability of AI is highly variable across SSA subregions, species and production systems. South Africa and Kenya are, by far, the highest users of AI in SSA, with well-established AI field programmes in place, predominantly serving the dairy sub-sector. In Kenya, imported semen still accounts for a higher percentage of inseminations than locally produced semen, and field insemination is dominated by private sector inseminators (Makoni et al. 2015). Kenya also exports locally produced as well as imported semen to neighbouring countries—especially Tanzania, Uganda and Rwanda—in programmes in which AI is used to facilitate crossbreeding of indigenous cows using semen from high yielding exotic cattle with the goal of increasing productivity. In addition to Kenya, the use of AI in eastern African countries can be seen in Uganda, Tanzania, Ethiopia and, to some extent, Rwanda. Sudan has also used AI for decades, but the coverage (number of inseminations) is relatively small. AI was first introduced in Ethiopia in the 1930s but was interrupted by World War II and subsequently went through up and down swings. The National Artificial Insemination Center (NAIC) was established in 1984 to coordinate the overall AI establishment and operation throughout the country. In the last 5 years, with support from ILRI's Africa Dairy Genetic Gains (ADGG) project, AI is beginning to pick up in the country.

Other than its use in dairy cattle, AI is also used in SSA in other livestock species to varying degrees: pigs, at least in Kenya, Uganda and South Africa; poultry in Kenya (Smallholder Indigenous Chicken Improvement Programme); and goats in Kenya and South Africa. The use of AI in beef cattle has mainly been limited due to the difficulty in detecting cows in heat within large herds and where individual cows are handled only occasionally. Countries using AI differ in terms of purpose and scope or extent of use, e.g. research, isolated field experimental use, limited use in commercial farms or limited geographies in peri-urban areas, to wider industry use in dairy cows and use in species other than cattle.

In almost all SSA countries, AI provision is dominated by the public sector—even where semen source is a combination of locally produced and imported. In such countries, AI field delivery is subsidized by the government. For example, the FAO (2015) survey points out that in Botswana, Ethiopia and Lesotho, semen doses are provided to livestock keepers at subsidized prices. South Africa (see Box 5.1) and Kenya (Box 5.2) have significant private sector involvement in AI field delivery and are exceptions.

Box 5.1 Private Sector-Led Applications of Reproductive Technologies in South Africa

The commercial dairy sector in South Africa is the largest user of reproductive biotechnologies (especially AI) in SSA. Imported semen (mostly Holstein-Friesian), which is cheaper than nationally produced semen, is widely used. Genetic evaluations are conducted by breed societies to ensure high standards are maintained.

Over the past 10 years, the pig industry has moved towards the use of hybrid genetics and AI, which is provided by two companies.

Imported embryos have been used to increase the numbers of Boran and Senepol cattle in the country, with varying degrees of success.

Limited semen sexing and in vitro fertilization are done by a few registered private sector service operators.

Cloning (somatic cell nuclear transfer) has been limited to research, with one clone of a dairy cow (see Sect. 5.2.3) having been successfully produced.

The country currently has 32 registered reproduction centres that provide semen and embryo collection services, AI and embryo transfer in cattle, sheep, goats and horses. There are over 300 trained inseminators in the country registered as per the country's Animal Improvement Act of 1998. Some provide AI services to the smallholder sector, but most are either owners of commercial dairy farms or employed on such farms.

Source: FAO (2015)

The Public-Private Partnership for Artificial Insemination Delivery (PAID), funded by the Bill and Melinda Gates Foundation, is underway to revitalize AI delivery system in Tanzania and Ethiopia. Other initiatives to expand use of AI in dairy cattle are under way in Cameroon and Ghana through the World Bank-supported West African Agricultural Productivity Program (WAAPP). In addition to use in research programmes in Senegal, limited AI is used in peri-urban commercial dairy farms, especially around Dakar. In 2014, Mali launched a major AI programme aimed at creating a larger number of crossbred cattle for improved milk and beef production.

In Botswana and South Africa, AI is used in beef herds to avoid transmission of reproductive diseases by bulls. Some results in Botswana suggest that AI is cheaper than use of bulls in beef systems. The government-subsidized AI programme in Botswana also serves to introduce superior sire lines in herds of smaller producers.

The Bambui and Wakwa centres of the Institute of Agricultural Research for Development in Cameroon (Bayemi and Mbanya 2007) are the only functional AI centres in the Central Africa Region—serving Chad, Central African Republic,

Gabon, Equatorial Guinea and the Congo Republic. Semen from Holstein Friesian bulls is used on local cows to improve the dairy potential of the resulting crossbreds. In Nigeria, beyond isolated use by individual commercial farmers, the National Animal Production Research Institute's AI unit has a research and training mandate on AI, but this is not linked to any field delivery.

Government-driven AI service serving a small number of farmers in limited areas has been running in Malawi since mid-1980s. In 2017, the Rural Livelihoods and Economic Enhancement Programme, with funding from the International Fund for Agricultural Development (IFAD) and the government of Malawi, strengthened the AI service delivery around Blantyre and Thyolo.

Box 5.2 The Emergence of Private Sector Leadership in AI in Kenya

The number of dairy cows in Kenya is estimated at 3.5 million. Only 18% of the dairy herd and less than 0.05% of the beef herd is bred through AI. Recorded annual inseminations averaged about 1.7 million from 2006 to 2013.

This is projected to change significantly following changes in the AI value chain—especially the liberalization that has led to massive entry of private inseminators and increases in volume of direct imports of semen by farmers and genetics companies. As of 2015, there were at least 14 local and international companies involved in the bull semen business.

The liberalized environment has catalysed emergence of a competitive and sophisticated environment. On the demand side are the smallholder and commercial dairy farmers, dairy farmer cooperatives and self-help groups, commercial milk bulking enterprises and milk collection centres; on the supply side, the Kenya Animal Genetic Resources Centre produces local semen and is involved in distribution of semen and liquid nitrogen, private suppliers of breeding inputs (bull semen, liquid nitrogen, companion breeding supplies), breed associations and related organizations (Ayrshire, Boran, Guernsey, Holstein/Friesian and Jersey Associations), Kenya Livestock Breeders Association, Livestock Genetic Society of East Africa, East Africa Embryo Transfer Association, Kenya Association of Livestock Technicians, Kenya Farmers Association, Kenya Dairy Farmers Federation and Kenya National Dairy Association of Producers.

As shown in the graph, imported semen has been increasing substantially from an average of about 50,000 units in 2006 to more than 350,000 units in 2013, a response to the increased demand for milk. In addition, the privatization of AI services initiated in 1991 resulted in almost all inseminations being carried out by private artificial insemination service providers; by 2014, government inseminators' coverage had dropped to less than 4%. Projections for the next 10 years indicated an increase in inseminations that will also increase demand for imported and local semen.

(continued)

Box 5.2 (continued)

Annual increase for both local and imported semen demand was projected to range from about 6 to 9%. By 2023, it is envisioned that there will be two million inseminations annually. As part of its AI expansion strategy to respond to anticipated increase in semen demand, the Kenya government announced the establishment of a second bull station (at Endebess) in Western Kenya in 2018, with a potential bull capacity of 100 and equipped with a state-of-the-art semen production unit.

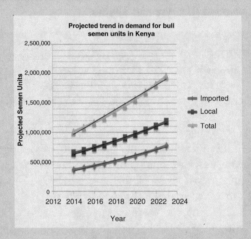

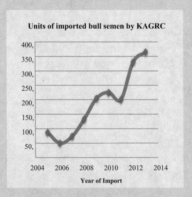

Source: Makoni et al. (2015) and compilation by authors

Oestrus Synchronization Oestrus synchronization (OS) is considered a companion technology to AI. The success of AI in the field depends on accurate detection of oestrus and skilful insemination, and OS allows a large number of females to be readied for insemination on a predetermined date. This reduces the inaccuracies associated with heat detection, increases conception rates and provides an opportunity to match calving period with feed availability.

Oestrus synchronization has been piloted widely across Africa and in multiple species, but the majority of cases have involved cattle. However, none of these pilots have evolved into an institutionalized practice integrated into a national livestock breeding programme. Perhaps the most extensive attempt conducted over an extended period of time was the mass on-farm OS undertaken in Ethiopia from 2007 to 2018 under two consecutive programmes—IPMS (www.ipms-ethiopia.org) and LIVES (www.lives-ethiopia.org) (Tegegne et al. 2016)—to produce a large crossbred dairy cattle population to kick-start market-oriented smallholder dairying. Even in combination with OS, lack of quality semen is a constraint. This is due to issues with handling and reliability of the cold chain in rural settings, including lack of and/or high cost of liquid nitrogen, quality of inseminations, poor road infrastructure and poor condition of animals associated primarily with nutrition.

Use of radioimmunoassay (RIA), a technique for measurement of the concentration of hormones, is very widespread across SSA for oestrus detection with predominant use being in dairy cattle, albeit mainly in research institutions, universities and diagnostics laboratories. The widespread availability and use of RIA have greatly benefitted from the efforts of the Joint FAO/IAEA Division of Nuclear Techniques in Food and Agriculture, which has assisted many member countries, including in SSA, to develop capacities for the use of nuclear and nuclear-related technologies in agriculture.

Technologies Applied to Animal Nutrition Technologies applied to animal nutrition include fermentation technologies used to produce nutrients, such as essential amino acids or complete proteins, or to improve the digestibility of animal feeds, and microbial cultures used to increase the quality of silage or to improve digestion when fed as probiotics. Recombinant bacteria, i.e. bacteria produced by genetic engineering, have also been developed that produce specific enzymes and hormones to improve nutrient utilization and increase productivity, e.g. commercial application of recombinant bovine somatotropin in South Africa, used to increase milk yield in dairy cows, or to decrease environmental impact, e.g. research trials with and commercial use of A B Vista's product Quantum Blue, an enhanced *E. coli* bacterial phytase in South Africa, which helps pigs and poultry to better utilize dietary phosphorus.

While most of the nutrition technologies are generally relatively simple, there are various factors which limit their use in developing countries. Silage production and use in SSA, for example, is not common, thus limiting the potential use of microbial silage cultures. The uptake of recombinant somatotropin has been affected by low public acceptance, the lack of adequate and good quality feed and the low genetic potential of the predominant livestock populations in SSA.

Probiotics and Prebiotics The term probiotics is used to describe live bacteria added to animal feed which have a beneficial effect on the host animal by affecting its gut flora. A prebiotic is a specialized plant fibre that beneficially nourishes the 'good' bacteria already in the large bowel or colon. Thus, while probiotics introduce 'good' bacteria into the gut, prebiotics act as a 'fertilizer' for the good bacteria that are already there.

Probiotics and prebiotics are increasingly used in commercial animal production operations to advantageously alter gastrointestinal flora, thereby improving animal nutrition, health and productivity. The major outcomes from using probiotics include improvement in growth, reduction in mortality and improvement in feed conversion efficiency.

Probiotics have been used specifically for prevention of bacterial diseases in poultry. Given its wide use in commercial poultry, the technology can be said to be available in all countries with significant commercial poultry farming. Probiotics are not used in smallholder systems, and the research community in many African countries is not aware that probiotic usage is widespread in commercial operations in their countries.

Among SSA countries, South Africa has had the highest rate of increase in probiotics usage in recent years, mainly in dairy cattle, pigs and poultry. The growth is due to increased awareness among farmers about the positive effects of probiotics on animal health and growth performance.

Smallholder chicken production in SSA is predominantly free-range with limited, if any, supplementary feeding—let alone use of probiotic and similar technologies. However, there have been trials to examine potential benefits of probiotics in these systems, with varying results (Khobondo et al. 2015; Atela et al. 2015). Combinations of *Bacillus* probiotic and xylanase, amylase and protease enzymes in a specific necrotic enteritis challenge model in Nigeria have shown net benefits of 14% in relative cost per kilogram live weight gain, illustrating the economic value of this combination and its potential. Probiotic supplementation has also been tried in cattle and other ruminants with favourable results (e.g. Yunus 2016), but, other than South Africa, there is no commercial use of the technology in ruminants in SSA.

Major companies active in the African feed probiotics industry are BioGaia AB (Sweden), China-Biotics Inc. (China), Chr. Hansen A/S (Denmark), Danone SA (France) and E.I. du Pont de Nemours and Company (USA), among others.

Silage Additives Fermentation in the silo can be an uncontrolled process leading to suboptimal preservation of nutrients. Silage additives have been used to improve the ensiling process (better energy and dry matter recovery) resulting into improvements in animal performance. A silage additive is a substance containing bacteria and/or chemicals, used to speed up or improve the fermentation process in silage or to increase the amount of nutrients in it. Similar to the pattern in probiotics, this technology is generally available and used in commercial ruminant livestock operations across SSA, but its usage in smallholder systems is restricted by a combination of awareness and cost.

5.2.2   Medium-Tech Applications

Embryo Transfer   Embryo transfer (ET) is a process whereby embryos produced by fertilization in another animal or in vitro are collected and implanted into a receptive surrogate female which then carries the resulting foetus to term. ET offers some of the advantages of AI albeit at lower scale. For example, ET allows the production of many more progeny by superior females than they would do through their natural reproductive cycle. This is especially so when ET is coupled with ultrasound-aided ovum picking, followed by in vitro maturation and fertilization (IVF). ET also allows dispersal of genetics over long distances, especially where pure breeds are desired. Furthermore, embryos can be collected and frozen in liquid nitrogen for use at a later date or for in vitro genetic conservation. In vitro fertilization and ET are requisite technologies for performing genetic modification of livestock, especially using genome editing.

Embryo transfer is not in wide use in SSA countries, except in South Africa and Kenya. However, the technology is used to a limited extent in research institutions and veterinary colleges in some other East African countries, notably Uganda and Tanzania.

South Africa undertakes embryo collection and transfer routinely, especially for cattle, sheep and the Boer goat breed, and some embryos are exported. The International Embryo Transfer Society (IETS) collects data on production and transfer of embryos in participating member countries. The IETS 2019 report presents data only from South Africa, although their earlier reports include data from Kenya and Namibia. These data point to three things: firstly, application of embryo transfer technology remains limited; secondly, while Kenya and South Africa are the major players in Africa, South Africa is clearly leading in terms of collection, domestic embryo transfers and exports, and indeed the country is ahead of many countries in the more developed regions of the world in this regard; and finally, while the potential for the exploitation of the technology exists, SSA lags behind other continents based on current statistics.

Commercial famers in East Africa and South Africa use the technology as a means of importing exotic germplasm of different livestock species, mostly cattle. Embryo transfer is being established in Cameroon as a tool to import high yielding dairy cattle genetics. Other countries which have varying levels of ET research work include Namibia, Côte d'Ivoire, Zambia and Zimbabwe. Generally, only a limited number of farmers in SSA are able to pay for the high-value genetics marketed through ET. This, in large part, explains the limited use to date.

The use of multiple ovulation and embryo transfer (MOET) allows cows of high genetic potential to produce a much larger number of calves than is possible with normal reproduction. Most MOET schemes require one or a few large nucleus herds. The resulting genetic improvement can be disseminated to the general population by ET and AI or by young bulls through natural breeding. Experience from South Africa shows that MOET can be practised in specialized centres and embryos transferred to other countries. This allows embryos produced by selected parents to be exported either within or outside the country of origin, thus shortening the

Table 5.3 Status of use of semen sexing and in vitro fertilization technologies

Region	No. of countries responding	Number of countries reporting use of:	
		Semen sexing	In vitro fertilization
Eastern Africa	7	4	1
Southern Africa	12	1	0
West and North Africa	19	1	1
Africa	38	6	2
World	120	48	47

Source: FAO (2015)

generation interval compared to use of AI. Even when import bans are in place precluding imports of animals, e.g. due to bovine spongiform encephalopathy (BSE—'mad cow disease'), embryos of highly productive parents can often still be imported, thus speeding genetic gains. Similarly, more productive breeds in one country can be transferred to another if the environment is similar (e.g. importation of Boran embryos from East Africa by Cameroon).

Sexing Sperm (Seidel et al. 1999) and embryo (van Vliet et al. 1989) sexing allow selective production of the desired sex of progeny to suit the production objectives of farmers, for example, production of heifers in dairy herds. Sexed semen is now commercially available, and there are reported cases of use in Uganda, Kenya, Tanzania and South Africa.

ILRI, working in collaboration with the Brazilian Agricultural Research Corporation (EMBRAPA), has conducted experiments aimed at introducing semen sexing into dairy systems in East Africa and, in collaboration with Kenyan scientists, has conducted an economic analysis which points to high potential for the combined technology—sexed in vitro fertilization embryo transfer (SIFET)—in Kenya (Lawrence et al. 2015).

Use of ovum pick up from live animals and sexed semen in sexed in vitro embryo production and embryo transfer, which involves both in vitro embryo production (IVEP) and ET, has also been demonstrated in research in Kenya and shown to have potential for delivering appropriate cattle genotypes to farmers (Mutembei et al. 2016). However, only South Africa has a significant commercial service of sexed embryos, for example, EmbryoPlus, a specialist breeding company established in 1980. Widespread use of sexed semen technology on smallholder farms is still limited by the inherent infrastructural constraints of AI delivery in most SSA countries and the relatively high cost. Nonetheless, the technology offers advantages and has prospects for adoption in countries with functional AI infrastructure.

Table 5.3 presents results of a 2014 FAO survey (FAO 2015) on status of use of semen sexing and IVF technologies in livestock. Although the number of responding countries was low (only 38 from Africa and a total of 120 globally), with regard to medium-level technologies in SSA, the picture reflects the status quite accurately. Four countries in eastern Africa (Kenya, Uganda, Rwanda and Tanzania) used sexed semen, with Kenya having also conducted some research on the technology; the other countries are mainly users of commercial embryos imported from different sources. In eastern Africa, Kenya is the only country using in vitro fertilization, and South Africa is the only country using the technology in the southern Africa region. North and West African countries are grouped together in the survey summary, and only one country (from North Africa) was reported to be using semen sexing and in vitro fertilization.

Kenya and South Africa are increasing their depth and scope of exploitation of these technologies with active involvement of the private sector. A few other countries, mostly dependent on public sector support of the technologies (in research and, specifically for AI, in provision of service to commercial operations in high potential or peri-urban areas), are continuing to do so but at low and varying levels—often 'on-and-off'—depending on their ability to address implementation challenges. There do not appear to be any recent entrants into field applications of these technologies over the last decade or so by those countries which had not ventured into these technologies before.

Cryopreservation of Germplasm   Conservation of indigenous genetic resources is one of the top priorities of developing countries, and demand for establishing cryopreservation was expressed in the FAO country reports of 2014 (FAO 2015). Ex situ conservation is considered a critical complement to in situ conservation. However, unlike in crops, cryoconservation in liquid nitrogen represents the only real solution in this regard for now.

In line with the Global Plan of Action for Animal Genetic Resources, the implementation of which is being coordinated by FAO, the African Union-Inter-African Bureau for Animal Resources (AU-IBAR) developed a strategy for the use of animal resources gene banks. The five planned gene banks were officially launched in 2019: for eastern Africa, situated at the National Animal Genetic Resource Centre and Data Bank, Entebbe, Uganda; northern Africa, located at the National Gene Bank of Tunis, Tunisia; western Africa, based at Centre International de Recherche-Développement sur l'Elevageen zone Subhumide (CIRDES), Bobo-Dioulasso, Burkina Faso; southern Africa, situated at the Department of Agricultural Research, Gaberone, Botswana; and Central Africa, housed at the University of Dschang, Cameroon. Although these gene banks have been officially launched, they are yet to be operationalized. It has been proposed that AU-PANVAC hosted in Debre Zeit, Ethiopia, which has been in existence since 2004, will hold backup samples of each region, thus providing security against accidental loss. These facilities also provide the host countries additional access to opportunities for engagement in aspects of livestock biotechnology.

At present, some of the well-established AI facilities in a few SSA countries are also being used for long-term preservation of semen of endangered or priority national animal breed resources. Examples include Kenya, Ethiopia, Uganda and South Africa. In some of these facilities (e.g. South Africa and Kenya), preservation of embryos is also being practised or being explored. Since 2011, In Vitro Africa, a

Table 5.4 Status of research in the use of molecular technologies in livestock[a]

Region	Number of responding countries	Number of responding countries reporting research on:							
		Adaptive traits		Genetic diversity estimation		Breeding value prediction		Genetic modification	
		Na	Int.	Na	Int.	Na	Int.	Na	Int.
Eastern Africa	8	2	2	4	3	1	1	0	0
Southern Africa	12	1	2	2	1	1	6	0	0
West and North Africa	20	4	3	7	6	24	3	0	0
Africa	40	7	7	13	10	6	7	0	0
World	128	47	36	70	55	52	46	22	15

[a]Note: Na = national institutions; Int. = international institutions; source: FAO (2015)

private company in South Africa, established facilities for the preservation and dissemination of embryos and semen.

Technologies for somatic cell cloning are developing rapidly and supports the case made over a decade ago (e.g. Hodges 2005) that samples of cryopreserved cells of all breeds should be stored long term in secure locations to be accessed if and when the need arises in the future. This could be either to sequence their DNA to understand genetic differences among breeds or to use the cells in cloning to regenerate extinct breeds.

Molecular Markers The advent of molecular biology in the 1970s, notably the discovery of restriction enzymes, invention of DNA sequencing (Sanger et al. 1978) and emergence of recombinant DNA technology, ushered in DNA-based polymorphism (DNA markers) as tools for genetic improvement of livestock. In Africa, the availability of molecular markers has triggered interest in genetic diversity studies and created a momentum in this area. An indication of this is seen in Table 5.4, which is a summary of status of research on molecular marker technologies in Africa in 2014 (FAO 2015). In addition to research on breed diversity characterization (reported by 4 out of 8 countries in eastern, 2 out of 12 in southern and 7 out of 20 in North and West Africa), markers are being used in studies of adaptive traits by national and international institutions. Indeed, at continental level, the reported engagement of national and international institutions in research involving marker technologies is almost at par. One country each in eastern (Kenya) and southern Africa (South Africa) reported use of markers in breeding value (BV) prediction research. The four countries reporting use of markers in BV prediction in the North and West Africa group did not include any SSA country. However, more recent information suggests that all countries in eastern Africa have been involved in one or more aspects of genetic diversity estimation using marker technologies.

One of the pioneering and well-documented applications of genetic marker technology in SSA was in the study of the trypanotolerance trait in N'Dama cattle with a view to developing specific genetic markers for the introgression of the trait into susceptible higher yielding breeds like the Boran (Hanotte et al. 2003).

The other major use of molecular makers in Africa has been in the genetic characterization of the diversity of African animal genetic resources (AnGR). The first continent-wide effort to characterize African AnGR was spearheaded by the then International Livestock Centre for Africa starting in 1992 (Rege and Lipner 1992), and this set the stage for a continental AnGR characterization effort covering all the major livestock species. Shortly thereafter in 1993, an FAO working group proposed a global programme for characterization of AnGR, including molecular genetic characterization, with recommendations for the molecular analysis of domestic animal diversity via a global research programme coordinated by FAO (FAO 1993).

These early major efforts have catalysed significant investments in AnGR characterization in Africa, covering several species—chickens (van Marle-Koster and Nel 2000; Mwacharo et al. 2011; Wragg et al. 2012), sheep (Muigai et al. 2009), goats (Chenyambuga et al. 2004), cattle (Okomo et al. 1998) and camels (Mburu et al. 2003)—and applying a range of markers, starting with the first-generation DNA markers (restriction fragment length polymorphism), minisatellites, microsatellites and mitochondrial D-loop DNA sequences. Today, the study of genetic diversity of livestock at the molecular level has developed into an active area of research, with SSA results receiving considerable attention in the scientific press and at continental and global conferences, with many such studies emanating from graduate student research.

The state of use of advanced DNA markers, such as high-density single nucleotide polymorphism (SNP), and genome sequencing for AnGR characterization is limited to a few populations/breeds of cattle, sheep, goats and chickens from Kenya, Uganda, Guinea, Ethiopia, South Africa and Nigeria.

AU-IBAR, in collaboration with FAO and ILRI, is currently (2020) working with African countries in undertaking an extensive project to characterize all major domestic species using genome sequencing technology. In addition, there are species-specific projects under way which are looking at both breed characterization and enhanced utilization—including genomic selection—in various species: camels (Mwacharo personal communication), goats (African Goat Improvement Network supported by USAID Feed the Future, in which SNP genotyping is being used to characterize goat populations in Kenya, Uganda, Tanzania, Ethiopia, Sudan, Malawi, Mozambique, South Africa, Zimbabwe and Nigeria—AGIN 2016) and chickens in Tanzania, Ethiopia and Nigeria (ACGG, https://africacgg.net/). An ILRI-led project, concluded in 2012, used high-density SNP technology to characterize, in order to optimize choice among, different crossbred dairy cattle in smallholder systems in Kenya and Uganda (Ojango et al. 2014).

DNA markers have also been used in SSA for pedigree ascertainment, individual animal identification, breed ascertainment and MAS or genomic breeding. Examples of these from South Africa include genome-wide scan for signatures of selection in cattle (Makina et al. 2015) and DNA-based parentage verification and diagnostic testing (Van Marle-Köster and Nel 2003). These examples illustrate potential and actual applications beyond research, and various projects in NARIs and universities in SSA are increasingly using these technologies and tools. Prospects and outlook for use of DNA markers in practical animal genetic improvement are good.

ILRI, in its facilities in Kenya, has a genomics platform, which also hosts a biorepository (biobank), the ILRI Azizi Facility, containing over 450,000 samples that include blood, tissues and semen, among others, from livestock, wildlife, humans and insects collected from East, West and Central Africa. Outputs from the facility include whole- or partial-genome sequences of pathogens, e.g. Rift Valley fever, equine encephalosis, blue tongue, African swine fever and Newcastle disease.

Diagnostic and Vaccine Technologies Diagnosis of disease is traditionally done by clinical examination accompanied by direct demonstration of the disease-causing agent, by either culture or microscopy, or the responses to infection through serology. Modern biotechnology has provided more robust tools which can facilitate diagnosis and vaccine development.

ELISA Enzyme linked immunosorbent assay (ELISA) is a biotechnology that revolutionized diagnosis through increased accuracy and sensitivity of detection of both antigens and antibodies. ELISA can be adapted to detect both antibodies from a past exposure and current infection. Nearly all SSA countries have laboratories equipped to carry out ELISA either for research, epidemiological surveys or field use in farmers' flocks and herds. ELISA tests are available for all the major diseases, including foot-and-mouth disease (FMD) in ruminants and pigs, pest de petit ruminants in sheep and goats, Newcastle disease in poultry, African swine fever in pigs and various tick-borne diseases. The current state of use of ELISA technology, while widespread across countries, is, to a large extent, limited to government diagnostics laboratories, some public livestock research centres and universities.

DNA Markers and Polymerase Chain Reaction The polymerase chain reaction (PCR) is a laboratory technique used to make multiple copies of a segment of DNA to facilitate its utilization as a molecular marker. PCR is very precise and can be used to amplify or copy a specific DNA target from a mixture of DNA molecules. PCR has found widespread uses: to diagnose genetic diseases, for DNA fingerprinting, for disease diagnosis (i.e. to find existing bacteria or viruses in samples), to study animal evolution, to clone the DNA of an ancient biological sample and to establish paternity or biological relationships, among others. Accordingly, beyond its common use in diagnostics, PCR has become an essential tool for biologists across several fields, such as DNA forensic labs and many other laboratories that study biological material.

Today, there are PCR tests available for use to diagnose FMD, ASF, animal African trypanosomiasis and various tick-borne diseases. As with ELISA, PCR diagnosis is available in almost all countries with veterinary laboratories, but it is mainly confined to laboratory settings in the diagnostic labs as well as in research institutions. The outbreak of avian influenza in 2005 in Africa, and the interventions that followed in response, expanded the capacity to do PCR in Africa as, with international development support, many more labs were equipped to deal with the outbreak.

Development and Application of Vaccine Biotechnology   Most veterinary vaccines used in Africa are inactivated agents (killed) or attenuated live agents. Modern vaccines can be prepared using recombinant DNA technology to identify, isolate and produce the protective antigen in large quantities either as protein, or in the case of recombinant vector vaccines, to use a harmless virus or bacterium as a vector, or carrier, to introduce genetic material associated with the protective antigen.

All African countries use vaccines to control important diseases with vaccines against Newcastle disease, FMD and pest de petit ruminants being the most commonly used across the continent. Other vaccines used include fowl pox, fowl typhoid and Gumboro disease in poultry, capri pox, black quarter and anthrax. While these vaccines are generally available in SSA countries, access and usage by farmers vary greatly among countries. Most of the countries do not have domestic capacities to produce the vaccines they need and use: there are vaccine production laboratories in just 11 SSA countries.

Poultry vaccines are routinely used by farmers with commercial poultry breeds but rarely in backyard systems based on indigenous chickens. In Tanzania, it is reported that only 22% of farmers regularly vaccinate (Tanzanian Ministry of Agriculture 2012). A project undertaken by WAAPP in Ghana promoted the use of Newcastle vaccination in backyard chickens and guinea fowls and aimed to vaccinate 65% of the national flock of 19 million birds kept by smallholder farmers by the end of the project in 2017.

With the exception of the recombinant rinderpest vaccine used in the Pan African Rinderpest Campaign to successfully eradiate rinderpest, and some vaccines used in the commercial poultry sector, most livestock vaccines currently used on the continent are conventional vaccines—live or inactivated. In the case of East coast fever (ECF), a tick-borne protozoan disease of cattle in eastern, Central and southern Africa, vaccination entails concurrent inoculation of the native virulent pathogens and antibiotic to limit progression of infection, known as the infection and treatment method (ITM) (Perry 2016). This is based on a protocol developed more than 50 years ago by the then East Africa Agriculture and Forestry Research Organization at Muguga, Kenya. The ECF ITM story is presented in Sect. 5.4.1. The best-documented case of use is in northern Tanzania where the ITM has been adopted by Maasai pastoralists and has resulted in reduction in calf mortality by up to 90% (Di Giulio et al. 2009).

5.2.3 High-Tech Applications

Somatic cell nuclear transfer (SCNT) is a process by which the nucleus of somatic cells is implanted into a recipient ovum shell, the nucleus of which has been extirpated, and stimulated to dedifferentiate and eventually grow into a new individual which is identical to the individual from which the nuclear was obtained. Somatic nuclear transfer offers some possibilities such as replicating an individual with very special attributes, a route for genetic modification and cryoconservation where embryos or gametes are not available. The global success rate of producing live cloned offspring from high-quality domestic livestock has improved in the past few decades, since the successful cloning of Dolly the sheep in 1996 (Wilmut et al. 1997).

There are only two reports of successful SCNT in cattle in SSA. The first cloned animal on the continent was a Holstein heifer named *Futhi* (Zulu for replica), born at the AI Centre at Brits in the North West Province of South Africa in April 2003. It was derived from a single cell taken by biopsy from the ear of a donor cow, inserted into an empty cow-egg and later implanted into a recipient cow. This work was a collaboration between scientists from South Africa and Denmark. The second cloned livestock was named *Tumaini* (Swahili for hope), a Kenyan Boran bull born on ILRI's research ranch in Kabete, Kenya. The animal was cloned by SCNT using primary embryonic fibroblasts. The ILRI team contends that this success opened the possibility of making genetically engineered cattle with foreign genes or desired traits through genome editing at the fibroblast level followed by SCNT, for example, the generation of a population of productive and disease-resistant/disease-tolerant livestock for specific contexts.

The high cost of cloning technology makes it not economically viable in commercial animal agriculture. It will continue to be used for research purposes and applied only under very special circumstances, e.g. when a critical genetic line is on the verge of extinction and conventional breeding is not an option.

Genetic modification is defined in the Cartagena protocol to the Convention on Biodiversity as the changing of the genetic makeup of an organism by artificial purposeful insertion, deletion and/or replacement of a segment of the DNA of a target organism. Genetic modification potentially offers unlimited opportunity to tailor livestock to the production environment in a way and pace that natural mutations or selective breeding cannot. Genetic modification of livestock can be used to enhance desirable traits or to eliminate undesirable traits in a few generations, allows importation of traits between species and can help overcome natural reproductive barriers.

Globally, there is only one example of genetically modified livestock that has been approved for commercial use. This is the ATrymGoat, developed by GTC Biotherapeutics to produce antithrombin in milk for use in the pharmaceutical industry, which was approved by the US Food and Drug Administration in 2009.

Genetic modification has become technically more tractable with the invention of gene editing. The only known piece of research in this line in SSA is that being pursued at ILRI to develop genetically modified cattle that are resistant to trypanosomosis. This is the basis of the work that led to the development of the Boran clone *Tumaini*.

5.3   Classification of Countries

Table 5.5 categorizes countries according to their biotechnology applications in livestock, ranging from 'very low' use to 'very high' use. No countries were categorized as having 'very high' use of biotechnology applications in livestock, and just two, Kenya and South Africa, were ranked as 'high'. A further 10 countries fell into the 'medium' use category, while the remaining 33 countries for which information was available were categorized as 'low' or 'very low'.

Overall, livestock comes second to crops in intensity of biotechnology applications. Several high-tech applications are being used in many countries although level of application is highly variable among countries, and it is also clear that far fewer countries have embraced these biotechnologies in livestock compared to crops.

5.4   Impact Cases in Livestock

This section provides four short case studies to demonstrate the potential of biotechnology applications in the livestock sector. The case examples are the case of infection and treatment method (ITM) as an immunization method that has been successfully applied in the control of ECF—a tick-borne disease of cattle—in eastern and southern Africa; the sterile insect technique (SIT) used in the eradication of the tsetse fly on an island in Tanzania; impact of applying farmer participatory breeding to improve the local Horro chicken in Ethiopia; and a successful

Table 5.5 Classification of countries on the basis of biotechnology applications in livestock

Applications category	Countries[a]
Very low use	Angola, Benin, Chad, Central African Republic, Republic of Congo, Djibouti, Equatorial Guinea, Eritrea, Eswatini, Gabon, Gambia, Guinea, Guinea Bissau, Lesotho, Liberia, Madagascar, Niger, Somalia, South Sudan, Togo
Low use	Botswana, Burkina Faso, Burundi, Cameroon, DRC, Malawi, Mozambique, Sierra Leone, Rwanda, Senegal, Zambia, Zimbabwe
Medium use	Côte d'Ivoire, Ethiopia, Ghana, Mali, Mauritius, Namibia, Nigeria, Sudan, Tanzania, Uganda
High use	Kenya, South Africa
Very high use	None

[a]Some countries are not listed due to lack of sufficient data: Cape Verde, Comoros, Mauritania, Sao Tome and Principe and Seychelles

demonstration, in sheep, of the application of a community-based breeding programme, also in Ethiopia.

5.4.1 Infection and Treatment Vaccine for East Coast Fever

East coast fever (ECF) is a devastating tick-borne disease of cattle caused by the protozoan parasite, *Theileria parva*. The disease causes high mortality, over 80%, in susceptible cattle populations. It occurs in 11 countries in eastern, southern and Central Africa where the tick vector, *Rhipicephalus appendiculatus*, is found, and causes major economic losses throughout the region, especially in crossbred and high-grade exotic dairy cattle and also young indigenous cattle in pastoralist systems and ranches.

The infection and treatment (ITM) immunization procedure involves the use of well-characterized live sporozoite forms of the *T. parva* parasites administered to cattle (the infection) simultaneously with a long-acting formulation of the antibiotic oxytetracycline (the treatment*)* (Perry 2016; Jumba et al. 2020). The result is an asymptomatic or mild episode of the disease followed by the animal's lifelong immunity to the disease. The ITM method was initially developed and refined at the former East African Veterinary Research Organisation at Muguga, Kenya, between 1967 and 1977. Now known as the Veterinary Research Centre, the institute is today part of the Kenya Agriculture and Livestock Research Organization (KALRO). Various versions of ITM immunization vaccine have been developed, each differing in the strains of *Theileria* used. The most widely used version, known as the Muguga cocktail, is a preparation of the Muguga strain and the Kiambu-5 strain isolates, derived from cattle *T. parva* and buffalo-derived *T. parva* strains (Peters et al. 2020).

Getting the vaccine out to the field had been constrained by safety concerns which remained unresolved for nearly three decades. In 1996, the International Livestock Research Institute (ILRI), at the request of the Food and Agriculture Organization of the United Nations (FAO), produced two commercial-scale batches of the Muguga cocktail, known as FAO-1 and FAO-2. In total, about 660,000 doses were manufactured and distributed on a commercial basis. While at this point the product had not been formally registered, based on the desperate need, its use was sanctioned by the Directors of Veterinary Services in the respective countries, pending formal registration. By 2006, the vaccine stock had been depleted. This was a clear sign of demand for the vaccine—especially considering that not all concerns about the product had been fully addressed and that, at USD8–12 per dose, the price was relatively high (Peters et al. 2020). Thus, about a decade later (in 2008) after lengthy consultations to address the underlying concerns, at the request from regional stakeholders under the auspices of the African Union-Inter-African Bureau for Animal Resources (AU-IBAR), ILRI produced a second batch.

By 2014, over one million cattle had been immunized with the ILRI-produced vaccine. With assistance from GALVmed, a not-for-profit organization that provides livestock treatments in developing countries, and with the support of donors, ILRI has worked with NARIs and directors of veterinary services in the region to register the vaccine in Kenya, Malawi and Tanzania. In Kenya, ILRI and KARI supported Kenya's director of veterinary services in trials that confirmed the safety and effectiveness of the Muguga cocktail-based vaccine, which was launched for national distribution in Kenya in December 2012.

Production of the live ECF vaccine is complicated, time-consuming and expensive. To produce one million doses requires 130 cattle that have not previously been exposed to the disease, 500 rabbits and at least 600,000 ticks. The entire process of making the batch takes up to 18 months. The product requires a cold chain and careful handling to deliver it and to have it administered by trained veterinarians in the field.

In 2014, the commercial production of the Muguga cocktail was taken up by the Centre for Ticks and Tick-borne Diseases (CTTBD), in Malawi, facilitated by GALVmed and the Bill and Melinda Gates Foundation. ILRI helped establish the vaccine production processes in Malawi, while GALVmed promoted commercial distribution. Efforts to register the vaccine in Uganda are ongoing, and four distributors in Kenya, two in Uganda and two in Tanzania are delivering the vaccine in these countries. The CTTBD is currently manufacturing three ITM vaccines, namely, the Muguga trivalent cocktail for eastern African countries, including Malawi; the Katete strain for Eastern Zambia; and the Chitongo strain for Southern Zambia.

ILRI is continuing research towards a better understanding of the biology of the disease, unravelling the immune mechanisms and developing molecular tools to characterize parasite strains, which will be important in ensuring the quality of future vaccine productions. In 2016, it was reported that 1.5 million doses of the Muguga cocktail of the ITM vaccine had been administered in 11 eastern and southern African countries. Use was especially high amongst cattle owned by Maasai herdsmen in northern Tanzania.

A study undertaken in Tanzania compared different ECF control options, including use of the ECF ITM vaccine and spraying with an acaricide to control ticks. It was concluded that vaccination combined with seasonal tick control was the most cost-effective control option. Another study in Tanzania demonstrated that use of the ECF ITM vaccine reduced the annual cost due to theileriosis by between 40 and 68% depending on the dipping strategy used in addition to vaccination. A further study undertaken in northern Tanzania, where ECF is reported to be responsible for more than 40% of annual cattle deaths, demonstrated that ITM was highly effective: calf mortality rates were reduced from more than 20% to around 2% per annum. However, it was found that wealthier cattle owners were more likely to use ITM than poorer ones; for example, while wealthier owners vaccinated up to 90% of their calves and immature animals, poorer owners vaccinated around 30%. Factors associated with these differences included the cost of vaccination, which was reported by the study to be about USD7 per animal, and the way the vaccine was packed. The ITM vaccine has to be stored in liquid nitrogen and used rapidly once thawed. It was packed in straws, each of which contained around 35 doses. This was convenient for cattle owners with large herds but less so for those with just a few animals. In this particular study, formal cost-benefit analysis was not done, but the study authors considered that ITM represented a highly positive outcome. In a different study in the Kenyan smallholder dairy sector, it was calculated that for every dollar spent on ITM, the cattle owner gained a benefit worth USD3.

The ECF ITM story demonstrates that African national agricultural research institutes can develop technologies with transformative potential. It also shows, however, that a biotechnology with great potential can remain unavailable to farmers not only because of technical hurdles, but also because of institutional and policy bottlenecks, and that getting the right partners and the partnership right is critical to delivery of technologies to the field. Finally, it shows that benefits from biotechnology may not be equally accessible and affordable to all members of society who need it.

For further information, see Nene et al. (2016).

5.4.2 Sterile Insect Technique for Tsetse Eradiation

Tsetse transmitted trypanosomes cause human African trypanosomosis (HAT, also known as sleeping sickness) in people and nagana in cattle, both of which are usually fatal if untreated. Thanks to improved diagnosis and better treatment options, the number of cases of HAT has reduced significantly over recent decades, but the cattle disease still causes huge losses: FAO estimates that the disease kills 3 million cattle a year across 37 African countries and causes economic losses of up to USD4 billion annually.

Between 1993 and 1997, a tsetse control campaign took place on Unguja Island, Zanzibar, with the ultimate objective of eradicating the tsetse species *Glossina austeni*. Although there are 22 species of tsetse across Africa, and usually multiple species exist together, Unguja had only 1 species. It was also an isolated island of around 1600 km^2 situated 35 km off the Tanzania coast. The cattle population of Unguja was reported to be 155,624 in a government census conducted in 2007/8.

The sterile insect technique (SIT) is a biological control method in which males of the target species are reared in captivity, sterilized and then released to compete with wild males.

Female tsetse mate only once and if they are mated by a sterile male they do not produce offspring. Normally, mated females go on to produce eight to ten larvae during their lifespan, which can be up to 4 months. Seven to 10 days after mating, female tsetse produce a single egg which develops into a larva in the uterus where it is nourished via 'milk' glands. The third instar larvae are deposited on the ground after about 9 days and they burrow underground to pupate, emerging as adult flies after around 30 days.

The SIT was developed in the 1950s and was successfully used first on an island off Venezuela and later in parts of mainland USA and Central America to eradicate screwworms, a type of fly, the larvae of which feed on the flesh of animals, including cattle.

Standard practice in the use of SIT was applied. So, initially, in 1993 the Unguja tsetse population was substantially reduced by the use of commercially available pour-on formulations of insecticide which were regularly applied to cattle in a programme implemented by the Zanzibar government and the United Nations Development Programme. This reduced the tsetse population by around 95%. Then, between 1994 and 1997, the SIT was implemented by the IAEA and the Tanzania government.

The Unguja SIT programme involved mass rearing of tsetse in colonies of almost one million females at a facility at Tanga, on the Tanzania mainland. The male offspring were irradiated in a gamma irradiator which renders them infertile but otherwise healthy. Batches of around 70,000 sterile males were then released on Unguja Island at weekly intervals. SIT depends on the sterile males outnumbering wild males and competing with them for wild females; in this case, the ratio of sterile to wild males was estimated by IAEA to be 10:1 when SIT was first introduced.

Large-scale releases of sterile males, dropped from light aircraft, began in May 1995. The last fertile female tsetse was captured in February 1996, no wild tsetse flies have been caught since September 1996 and the last case of nagana in cattle was diagnosed in August 1997. Unguja was declared tsetse-free in 1997, and in 2016 FAO celebrated 20 years since the last wild fly was caught on the island. In total, close to nine million sterile male tsetse were released. The cost of the SIT programme was estimated to be just under USD5.8 million. Independent estimates suggest it costs around USD0.10 to rear each sterile male.

Following successful eradication of tsetse on Unguja, farmers were spared the estimated USD2 million annual losses from nagana. Eradication also gave them confidence to invest in improved and more productive animals: in 2014, a survey revealed that the population of crossbred cattle had increased by 38% and these animals produced more than twice as much milk as the local breed.

Although tsetse flies were successfully eradicated from Unguja using SIT, the characteristics of the island differ from most of tsetse-infested mainland Africa. Unguja was a remote isolated island with a single species of tsetse. Prior to the SIT campaign, the tsetse population was substantially reduced: one independent analysis suggests that at the start of SIT campaign, the male tsetse population was as low as 1000 individuals, most of which were confined to a few square kilometres of the Jozani Forest. Further, this analysis suggests that given that it is necessary to substantially reduce the tsetse population before SIT can be deployed, it is likely that the approach used for this initial phase would have been capable of bringing about eradication had it been continued, and this may have been cheaper than the two-phase approach (insecticidal treated cattle followed by SIT) used in this case. In cases where more than one species of tsetse exists, the cost of rearing multiple species would substantially increase the cost of implementing SIT.

To overcome some of the disadvantages of SIT, an improved methodology known as 'boosted SIT' has been developed and tested on tsetse in laboratory conditions. This enhances conventional SIT by treating the irradiated sterile males with a powder formulation of a biocide, pyriproxyfen, prior to release. The sterile males transfer small amounts of pyriproxyfen to females during mating or mating attempts. Pyriproxyfen is a juvenile hormone analogue; in female tsetse, a dose of just 0.01 µg has been demonstrated to result in non-viable offspring for at least two reproductive cycles. Boosted SIT should, in theory, enable the number of sterile males that need to be released to be significantly reduced, which would make the approach more cost-effective. In the case of *Aedes* mosquitoes, it has been predicted that boosted SIT to control dengue epidemics could enable a 95% reduction in the number of sterile males needed.

5.4.3   The Horro Chicken Breed Improvement Programme

The Horro chicken breeding programme in Ethiopia was started in 2008 and implemented by ILRI in collaboration with Wageningen University (Nigussie et al. 2010). The objective was to improve production of village chickens through a participatory within-breed selection approach. The breeding objectives were identified using a participatory approach. The programme aimed to develop a dual-purpose chicken through selective breeding.

Initially, a survey involving 225 households was conducted to understand the production system, needs and constraints of smallholder chicken farmers. On the basis of the survey results, the breeding goal traits were determined as greater egg production and higher body weight, decreased age at first egg and enhanced survival.

The base population was established from 3000 eggs purchased from various locations in the Horro region of Ethiopia and taken to the Ethiopian Institute of Agricultural Research in Debre Zeit, Ethiopia. Twenty cockerels and 260 hens were hatched and raised, and these formed the parental population. Selection was based on individual performance until the eighth generation. Each generation of approximately 600 males and 600 females was produced as selection candidates and recorded for body weight and egg production. Females were selected based on performance for body weight and egg production. Males were selected based on their performance for body weight. Selection pressure was 10–20% in the males and 50–60% in the females.

Evaluation of the breeding program was conducted when the program was in the eighth generation. Breeding values were estimated both for cumulative egg number at 45 weeks of age and body weight at 16 weeks of age to evaluate the trend of changes over the generations. The genetic trends show that by the eighth generation, survival had improved from less than 50% in the base generation to almost 100%. Body weight at 16 weeks had increased substantially from 550 g to 1.1 kg. Egg production nearly tripled from 64 eggs per hen per year in the base generation to 172 by generation 8.

There has been a large and significant improvement compared to unselected village chickens as a result of the breeding program (Mulugeta et al. 2020). While the performance of Horro chicken is still lower than commercial lines, the difference is decreasing with each generation of selection, and the degree of improvement achieved to date is impressive. Indeed, when adaptability, taste and consumer preference are taken into account, the improved indigenous birds can be considered superior.

5.4.4 Community-Based Sheep and Goat Breeding Programme in Ethiopia

Community-based breeding programmes (CBBPs) for small ruminants have been suggested as an alternative to the centralized government-controlled breeding schemes which have been implemented in many developing countries. An innovative methodological framework on how to design, implement and sustain CBBPs was tested in three sites in Ethiopia and involving three breeds, the Bonga, Horro and Menz. In the CBBPs, selection traits identified through participatory approaches were 6-month weights in all the three sites, and in Horro and Bonga, where resources, particularly feed and water, permit larger litter sizes, twinning rate was included. Between 2009 and 2018, 10 years of performance data from the programme were analysed. Additionally, socio-economic impact of the CBBPs was assessed.

Substantial genetic gains were achieved, with 6-month weight, the major selection trait in the CBBPs, increasing over the years in all breeds. In Horro and Bonga sheep, where twinning rate was one of the selection traits, litter size increased over the years: by 15.4% in Bonga and 11.6% in Horro. A total of 3200 households in 40 villages had benefitted by 2019. These gains increased income by 20%, and farm-level meat consumption increased from slaughter of one sheep per year to three.

The results show that CBBPs are technically feasible, can result in measurable genetic gains in performance traits and can have significant impact on the livelihood of farmers.

Farmers have created 35 formal breeders' cooperatives to participate in the programme, and the approach has been replicated in more than 40 programmes which have sprung up in the country based on inspiration and learnings from the original sites.

Most of the participating households in Menz no longer need assistance from government-run safety-net programmes that provide food aid; they now use income from sheep sales to buy food. The breeding cooperatives have been able to build capital from buying rams and bucks, as well as from other investments. For example, the Bonga cooperative has a capital of about USD60,000. There has been a high demand to buy breeding rams and bucks from neighbouring communities and other governmental and NGO programmes. The government has identified CBBP as the strategy for genetic improvement of small ruminants in the Ethiopia Master Plan and Growth and Transformation Plan II, and the program is being replicated in Iran, Malawi, South Africa, Sudan, Tanzania and Uganda.

The Horro chicken (Sect. 5.4.3) and the community-based sheep and goat breeding programme of Ethiopia are demonstrations of opportunities available to develop functional community-based breeding programmes in Africa that can deliver within-breed genetic improvement. Lessons learnt from these and similar programmes can be used to inform improved design, including adjustments needed to ensure stronger community participation and ownership.

Source: Haile et al. (2019)

References

Atela JA, Tuitoek J, Onjoro PA et al (2015) Effects of probiotics feeding technology on weight gain of indigenous chicken in Kenya. IOSR J Agric Vet Sci 8(11):33–36

Baker RL, Lahlou-Kassi A, Rege JEO et al (1992) A review of genetic resistance to endoparasites in small ruminants and an outline of ILCA's research programme in this area. In: Proceedings of the 10th scientific workshop of the small ruminant collaborative research support program, Nairobi, Kenya, pp 79–104

Baker RL, Rege JEO, Tembely S et al (1998) Genetic resistance to gastrointestinal nematode parasites in some indigenous breeds of sheep and goats in East Africa. In: Proceedings of the 6th world congress on genetics applied to livestock production, vol 25, Armidale, pp 269–272

Baker RL, Nagda S, Rodriguez-Zas SL et al (2003) Resistance and resilience to gastro-intestinal nematode parasites and relationships with productivity of Red Maasai, Dorper and Red Maasai x Dorper crossbred lambs in the sub-humid tropics. Anim Sci 76:119–136

Bayemi PH, Mbanya NC (2007) The first cattle artificial insemination centre in Cameroon. IRAD scientific review conference, Yaoundé, July 2007, pp 3–5

Chenyambuga SW, Hanotte OO, Hirbo J et al (2004) Genetic characterization of indigenous goats of sub-saharan africa using microsatellite DNA markers. Asian-Australas J Anim Sci 17(4): 445–452

Di Giulio G, Lynen G, Morzaria S et al (2009) Live immunization against East Coast fever–current status. Trends Parasitol 25:85–92. https://doi.org/10.1016/j.pt.2008.11.007

FAO (1993) Secondary guidelines for development of national farm animal genetic resources management plans. Measurment of Domestic Animal Diversity (MoDOD): Recommended microsatellite markers. FAO, Rome

FAO (2015) The second report on the state of the world's animal genetic resources for food and agriculture, edited by Scherf BD, Pilling D. FAO Commission on Genetic Resources for Food and Agriculture Assessments. Rome. Available at http://www.fao.org/3/i4787e/i4787e.pdf

FAO (2016) Development of integrated multipurpose animal recording systems. FAO Animal Production and Health Guidelines, No. 19. FAO, Rome

Haile A, Gizaw S, Getachew T et al (2019) Community-based breeding programmes are a viable solution for Ethiopian small ruminant genetic improvement but require public and private investments. J Anim Breed Genet 136(5):319–328. https://doi.org/10.1111/jbg.12401

Hanotte O, Ronin Y, Agaba M et al (2003) Mapping of quantitative trait loci controlling trypanotolerance in a cross of tolerant West African N'Dama and susceptible East African boran cattle. PNAS 100:7443–7448

Hodges J (2005) Role of international organizations and funding agencies in promoting gene-based technologies in developing countries. In: Makkar HPS, Viljoen GJ (eds) Applications of gene-based technologies for improving animal production and health in developing countries. Springer Publishing Company, Cham, pp 18–21

Jumba H, Teufe N, Baltenweck I et al (2020) Use of the infection and treatment method in the control of East Coast fever in Kenya: Does gender matter for adoption and impact? Gend Technol Dev 24(3):297–313

Khobondo JO, Ogore PB, Atela JA (2015) The effects of dietary probiotics on natural IgM antibody titres of Kenyan indigenous chicken. Livest Res Rural Dev 27(11):27–30

Kosgey IS, Mbuku SM, Okeyo AM et al (2011) Institutional and organizational frameworks for dairy and beef cattle recording in Kenya: a review and opportunities for improvement. Anim Genet Resour 48:1–11

Lawrence FG, Mutembei H, Lagat J et al (2015) A cost-benefit analysis of usage of sexed in-vitro fertilization embryo transfer technology in Kenya. J Agric Sci Food Technol 1(4):53–58

Makina SO, Muchadeyi FC, van Marle-Köster E et al (2015) Genome wide scan for signatures of selection among six cattle breeds in South Africa. Genet Sel Evol 47(1):1–14

Makoni N, Hamudikuwanda H, Chatikobo P (2015) Market study on artificial insemination and vaccine production value chains in Kenya. Study commissioned by the embassy of the Kingdom of the Netherlands in April, 2015. Nairobi, Kenya. Available at https://agriprofocus.com/upload/Market_Study_on_AI_and_Vaccine_Production_Value_Chains_in_Kenya14298772 94.pdf

Malabo Montpellier Panel (2020) Meat, milk & more: policy innovations to shepherd inclusive and sustainable livestock systems in Africa, Malabo Montpellier Panel Report, 2020. Available at: https://www.mamopanel.org/resources/reports-and-briefings/meat-milk-more-policy-innovations-shepherd-inclusi/. Accessed on July 6, 2021

Matika OS, Nyoni JB, Van Wyk GJ (2003) Resistance of Sabi and Dorper ewes to gastro-intestinal nematode infections in an African semi-arid environment. Small Rumin Res 47:95–102

Mburu DN, Ochieng JW, Kuria SG et al (2003) Genetic diversity and relationships of indigenous Kenyan camel (Camelus dromedarius) populations: Implications for their classification. Anim Genet 34(1):26–32

Mosi RO (1984) Use of milk records in sire and cow evaluation in Kenya. PhD Thesis, University of Wales

Mrode R, Chinyere ED, Marshall K et al (2020) Phenomics and its potential impact on livestock development in low-income countries: innovative applications of emerging related digital technology. Anim Front 10(2):6–11. https://doi.org/10.1093/af/vfaa002

Muigai AW, Okeyo AM, Kwallah AK et al (2009) Characterization of sheep populations of Kenya using microsatellite markers: Implications for conservation and management of indigenous sheep populations. S Afr J Anim Sci 30(1):93–96

Mulugeta S, Goshu G, Esatu W (2020) Growth performance of DZ-white and improved Horro chicken breeds under different agro-ecological zones of Ethiopia. J Liv Sci 11:45–53. https://doi.org/10.33259/JLivestSci.2020.45-53

Mutembei HM, Mulei CM, Mbithi PM (2016) A Kenyan economic analysis on utilization of ovum pick up, in vitro embryo production and embryo transfer technologies in cattle. Int J Vet Sci 5(2):64–68

Mwacharo JM, Bjørnstad G, Mobegi V (2011) Mitochondrial DNA reveals multiple introductions of domestic chicken in East Africa. Mol Phylogenet Evol 58(2):374–382

Mwangi EK, Stevenson P, Ndung UJ (1998) Studies on host resistance to tick infestations among trypanotolerant Bos indicus cattle breeds in East Africa. Ann N Y Acad Sci 849:195–208

Nene V, Kiara H, Lacasta A et al (2016) The biology of Theileria parva and control of East Coast fever - crrent status and future trends. Ticks Tick-borne Dis 7(4):549–564. https://doi.org/10.1016/j.ttbdis.2016.02.001.Epub

Nigussie D, van der Waaij LH, Dessie T et al (2010) Production objectives and trait preferences of village poultry producers of Ethiopia: Implications for designing breeding schemes utilizing indigenous chicken genetic resources. Trop Anim Health Prod 42(7):1519–1529

Ojango JM, Marete AG, Mujibi, DF et al (2014) A novel use of high density SNP assays to optimize choice of different crossbred dairy cattle genotypes in small-holder systems in East Africa. Paper presented at the 10th world congress of genetics applied to livestock production. Vancouver

Okomo MA, Rege JE, Teale A et al (1998) Genetic characterisation of indigenous East African cattle breeds using microsatellite DNA markers. In: Proceedings of the 6th World congress of genetics applied to livestock production, 11 Jan 1998, vol 27, pp 243–246

Perry BD (2016) The control of East Coast fever of cattle by live parasite vaccination: a science-to-impact narrative. One Health 2:103–114

Peters A, Toye P, Spooner P et al (2020) Registration of the East Coast Fever infection and treatment method vaccine (Muguga cocktail) in East Africa [version 1; not peer reviewed].

Gates Open Res, 4:100. Available via Gates Open Res: https://doi.org/10.21955/gatesopenres. 1116653.1

Rasali D, Shrestha JNB, Crow GH (2006) Development of composite sheep breeds in the world: a review. Can J of Anim Sci 86:1–24

Rege JEO, Lipner ME (eds) (1992) African animal genetic resources: their characterization, conservation and utilization. In: Proceedings of the research planning workshop, 19–21 February. International Livestock Centre for Africa, Addis Ababa

Rege JEO, Tembely S, Mukasa-Mugerwa E et al (2002) Effect of breed and season on production and response to infections with gastro-intestinal nematode parasites in sheep in the highlands of Ethiopia. Livest Prod Sci 78:159–174

Rowlands GJ, Teale AJ (eds) (1994) Towards increased use of trypanotolerance: current research and future directions. Proceedings of a workshop organized by ILRAD and ILCA in April 1993. International Laboratory for Research on Animal Diseases/International Livestock Centre for Africa, Nairobi

Sanger F, Coulson AR, Friedmann T et al (1978) The nucleotide sequence of bacteriophage phiX174. J Mol Biol 125(2):225–246. https://doi.org/10.1016/0022-2836(78)90346-7

Seidel GE, Schenk JL, Herickhoff LA et al (1999) Insemination of heifers with sexed sperm. Theriogenology 52(8):1407–1420

Silva MV, Sonstegard TS, Hanotte O (2012) Identification of quantitative trait loci affecting resistance to gastrointestinal parasites in a double backcross population of Red Maasai and Dorper sheep. Anim Genet 43:63–71

Tanzania Ministry of Agriculture (2012) United Republic of Tanzania national sample census of agriculture: Livestock Sector

Tegegne A, Hoekstra D, Gebremedhin B et al (2016) History and experiences of hormonal oestrus synchronization and mass insemination of cattle for improved genetics in Ethiopia: from science to developmental impact. LIVES Working Paper 16. International Livestock Research Institute (ILRI), Nairobi

van Marle-Koster E, Nel LH (2000) Genetic characterization of native southern African chicken populations: evaluation and selection of polymorphic microsatellite markers. S Afr J Anim Sci 30(1):1–6

van Marle-Koster E, Nel LH (2003) Genetic markers and their application in livestock breeding in South Africa: a review. S Afr J Anim Sci 33(1):1–10

vanVliet VR, Gibbins AMV, Walton J (1989) Livestock embryo sexing: a review of current methods, with emphasis on Y-specific DNA probes. Theriogenology 32(3):421–438

Wambura PN, Gwakisa PS, Silayo RS et al (1998) Breed-associated resistance to tick infestation in Bos indicus and their crosses with Bos taurus. Vet Parasitol 77:63–70

Wilmut I, Schnieke AE, McWhir J et al (1997) Viable offspring derived from fetal and adult mammalian cells. Nature 385:810–813

World Bank (2014) Business and livelihoods in African livestock: Investments to overcome information gaps. World Bank, Washington, DC. https://openknowledge.worldbank.org/handle/10986/17801

Wragg D, Mwacharo JM, Alcalde JA et al (2012) Analysis of genome-wide structure, diversity and fine mapping of Mendelian traits in traditional and village chickens. Heredity 109(1):6–18

Yunus AA (2016) Effect of probiotic supplement on growth performance, diarrhoea incidence and blood parameters of n'dama calves. MSc Thesis, Kwame Nkrumah University of Science and Technology

The State of Capacities, Enabling Environment, Applications and Impacts of Biotechnology in the Forestry Sector

6

J. E. O. Rege and Joel W. Ochieng

Abstract

Sub-Saharan Africa countries were categorised with regard to their capacity, enabling environment and applications to date of biotechnology in forestry. Capacity assessment examined human capacities, institutions and facilities, operational budgets and existence of facilitating networks. Enabling environment analysis covered public awareness, participation and acceptance, existence of national biosafety frameworks, public policy, political goodwill and public and private investments. Research and application of biotechnology were considered based on both the extent and context of use. Other than South Africa which was categorized as 'high' and Ethiopia, Kenya and Nigeria which were categorized as 'medium', the remaining countries were categorized under either 'low' or 'very low' capacities. No country had a 'very strong' enabling environment for biotechnology applications in forestry. Ethiopia, Kenya, Nigeria, South Africa and Sudan were assessed as 'strong', while Botswana, Ghana, Malawi, Mali, Namibia, Tanzania, Uganda and Zimbabwe had 'medium' enabling environments. The remaining countries had either 'weak' or 'very weak' enabling environments for biotechnology applications in forestry. Out of the 44 countries with information, there was no country in the 'very high' applications/use, only South Africa was in the 'high' use category, 8 were in 'medium', and the rest (12) were 'low' or 'very low' (23).

J. E. O. Rege (✉)
Emerge Centre for Innovations-Africa, Nairobi, Kenya
e-mail: ed.rege@emerge-africa.org

J. W. Ochieng
Agricultural Biotechnology Programme, University of Nairobi, Nairobi, Kenya

6.1   Sub-Saharan Africa Forestry Sector

Africa is home to 30% of the world's remaining rainforest, and close to 23% of the African continent is covered by diverse forests, ranging from the dry forests of the Sahel and eastern, southern and northern Africa to the humid tropical forest in central and western Africa (Malhi et al. 2013). The continent also harbours vast areas of woodlands and savannas that are not systematically documented. Five countries with the largest forest area—the Democratic Republic of Congo, Sudan, Angola, Zambia and Mozambique—together account for about 55% of the continent's forests. Planted forests account for a total of 15.4 million hectares with the bulk being in North Africa, outside of SSA.

Two of Africa's forested areas, the Upper Guinea Forest of West Africa and the Eastern Arc Mountain Forest in East Africa, are recognized biodiversity hotspots, while the Congo Basin is the second largest contiguous expanse of tropical rainforest in the world and is home to 65% of biodiversity in SSA. The Democratic Republic of the Congo (126 million hectares of forest) was among the ten countries in the world with the largest forest areas, while Madagascar is among the top ten countries with the most tree species.

More than 90% of people living in extreme poverty depend on forests and woodlands for at least part of their livelihoods. On average, just under a quarter of household incomes in developing countries come from forests. In SSA, 70% of the population rely on fuelwood for their primary energy source with average per capita consumption estimated in 2011 to be around 0.7 cubic metres, or 2.5 times the global average (Sola et al. 2017). Estimates of the contribution of forest-related activities to GDP in SSA vary; for example, a review published in 2010 revealed that although this was estimated to average just 3% in many African countries, if the value of industrial wood products, ecotourism and non-wood forestry products was comprehensively included, this would be close to 20%.

The latest report on state of the world's forests (FAO 2020) states that deforestation and forest degradation continue to take place at alarming rates, which contributes significantly to the ongoing loss of biodiversity. Africa had the highest net loss of forest area in 2010–2020, with a loss of 3.94 million hectares per year, followed by South America with 2.60 million hectares per year. Since 1990, Africa has reported an increase in the rate of net loss, while South America's losses have decreased substantially, more than halving since 2010 relative to the previous decade.

Agricultural expansion continues to be the main driver of deforestation and forest fragmentation and the associated loss of forest biodiversity, with large-scale commercial agriculture, primarily livestock ranching and cultivation of soya bean and oil palm, having accounted for 40% of tropical deforestation between 2000 and 2010 and local subsistence agriculture for another 33% (FAO 2020). Of note is that most forest habitats in temperate regions support relatively few animal and tree species and species that tend to have large geographical distributions, while the montane forests, including those in Africa, have many species with small geographical distributions.

The conservation and sustainable management of forests within an integrated landscape approach that balances local needs and demands are considered key to the conservation of the world's biodiversity and to food security and well-being of the world's people (FAO 2020). There is need to ensure that biodiversity conservation is mainstreamed into forest management practices in all forest types. Biotechnology has a role to play in helping SSA in its efforts to conserve remaining forests and to achieve its reafforestation objectives. This is in addition to the indirect, but critical, role that biotechnology can increasingly play in facilitating the production of more food on less land, thus reducing the land expansion that continues to characterize SSA agriculture, with implications for deforestation.

In respect to biotechnology applications, the forestry sector is different from the crop or livestock sectors in many ways—all of which define areas of growing interest by forest scientists, conservationists and tree growers in modern biotechnology applications. Forest trees are primarily long-lived perennials. Trees are also characterized by high levels of heterogeneity, late sexual maturity and a lengthy regeneration cycle that ensures a high retention of genetic diversity as a safeguard against rapid changes. Therefore, development of improved varieties through biotechnology or others means can be a protracted process. Furthermore, most forest tree species have narrow regional adaptation, and therefore the number of species used for planting is considerably higher than for food crops. This makes regional upscaling of varieties improved through biotechnology more difficult. Forest trees also serve as keystone species in dynamic ecosystems, and their management to prevent losses takes more than tree survival. Moreover, forest trees are also largely undomesticated although a few species have benefitted from population-level improvement for one to four generations (FAO 2011a). Application of biotechnologies in forests has been seen as opportunity for unravelling and utilizing some of these unique features of trees to obtain new information on the extent, patterns and functioning of tree genetic diversity and to provide new tree varieties and reproductive materials adapted to changing environmental, social and economic environments.

6.2 Capacities for Biotechnology Applications in Forestry

Most SSA countries have defined their forestry work to focus on 'conservation and use' with emphasis on protection and rehabilitation/reafforestation. Unfortunately, research—including that which could support conservation and use—is not embedded in forest programmes of most countries. As a result, the capacity available for forest biotechnology research and applications is extremely limited in most countries—even more so than for livestock. Thus, although education and skills development are challenges for all the four sectors (crops, livestock, forestry and aquaculture), the immediate concern with food security has tended to bias the limited national investments to crops (and to a smaller extent livestock), with stronger emphasis on the use of new technologies including biosciences, in comparison to forestry. Capacities for research and development in forestry are, therefore, quite limited. Indeed, only a few SSA countries have established functional national institutions with forestry R&D remits and capacities to implement programmes. In most countries, the only visible R&D interventions in forestry are those being done through international partnerships and collaborations involving continental non-governmental networks of the limited numbers of forest professionals, including the significant work of the CGIAR centres, ICRAF and CIFOR, which focus on research but also make significant contributions to development, including policy.

It is also noteworthy that, unlike crops, livestock and fisheries, in which private sector players are active, albeit to variable extents across countries, the private forestry sector in many countries continues to be almost non-existent, being made up of many small, dispersed and unorganized actors, who lack resources for investment and have limited voice (AFF 2019). In addition, forestry does not appear in many SSA government plans, nor in allocation of national resources. The absence of this important stakeholder category and the innovations it brings represents a major gap.

The limited infrastructure and facilities for biotechnology R&D are mainly located in national forestry research institutes and forestry schools or departments of public universities, where these exist. Moreover, available facilities are mostly for low-tech applications—mainly tissue culture and production of bio-fertilizers. A few countries have in the recent past acquired minimal capacities for medium-level biotechnology applications, such as use of molecular markers in phylogeny, molecular characterization and genetic diversity studies.

6.2.1   Human Resources

Figure 6.1 presents the state of human resources available for the forestry sector of SSA countries that have data on the ASTI platform. The figures show that, although human resources capacity in forestry biotechnology in SSA is generally low, there are important country differences. Comparable quantitative data on South Africa was not available. However, as indicated in the section below, the country has significant capacities and activities in forestry. Unsurprisingly, countries that ranked highest in terms of human capacity in the crop and livestock sectors are also in the top ten with respect to forestry, i.e. Ethiopia, Kenya, Niger, Ghana, Nigeria, Burkina Faso, Tanzania, Zambia, Madagascar and Zimbabwe.

Available FTE data could not be unpacked to quantify the biotech content as they are generally reported as 'forestry experts' and not disaggregated into disciplinary clusters (e.g. genetics, dendrology, forest ecology, silvics, wood structure and properties, forest soils, forest entomology and forest pathology).

6.2.2   Institutions and Facilities

Only a small number of SSA countries have dedicated forestry institutions and facilities. Some countries without dedicated forestry institutions have established

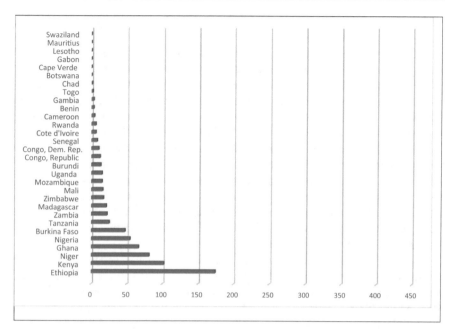

Fig. 6.1 Total forestry FTEs in SSA countries (2014)

forestry R&D as units or programmes within their agricultural research institutions (NARIs) and universities.

In Madagascar, capacities for micro-propagation of disease-free tropical trees are available in the National Center for Applied Research on Rural Development (FOFIFA) and in the University of Antananarivo, both of which are supported by CIRAD, France. Tanzania has a forestry institution—the Tanzania Forestry Research Institute—which has facilities for tissue culture applications. Forestry in Botswana is a sub-sector under the Ministry of Agriculture and does not include an explicit research mandate. However, the University of Botswana has some laboratory facilities for micro-propagation of forestry species. The Forestry Research Institute of Ghana (FORIG) has limited human and infrastructural capacities to conduct tissue culture and some basic molecular marker technology for research and applications. In Sudan, the Forest Research Center operating under the ARC has a tissue culture laboratory and some molecular biology facilities to undertake minimal marker technology work (Ali 2009). The combined human and infrastructural capacity available in the Forestry Research Institute of Malawi (FORIM), Lilongwe University of Agriculture and Natural Resources (LUANAR) and University of Malawi allows them to conduct low-level and some medium-level biotechnology research and applications. In Zimbabwe, human and infrastructural capacities are mainly in the University of Zimbabwe and at the Forestry Commission which, between them, conduct research and applications in low-level technologies, such as tissue culture, as well as medium-level technologies, mainly forestry genetic diversity studies (Olembo et al. 2010).

The Kenya Forestry Research Institute (KEFRI) has a modern forestry biotechnology laboratory with sufficient human and infrastructural capacity to conduct tissue culture and molecular marker technology research and applications—mainly molecular characterization and genetic diversity studies in forestry species. In addition, the Industrial Crops Research Institute, as well as other institutes working on tree crops such as the Coffee Research Institute and Tea Research Institute, all operating under KALRO, has the requisite human and infrastructural capacities for tissue culture and molecular marker technology for their mandate tree crops. The laboratories in these institutions also have capacities for some genetic engineering, particularly genetic modification using somatic embryogenesis. Tertiary training in forestry is undertaken in specialist training colleges, most of which focus on lower cadre (certificate and diploma level) field and operational personnel with limited capacity to contribute to biotech research and applications.

The Forestry Research Institute of Nigeria (FRIN) has specialized research departments covering Sustainable Forest Management, Forest Product Development and Utilization, Forest Conservation and Protection, Forest Economics and Extension, Environmental Modelling and Management and Wildlife and Tourism. Their research focus includes biodiversity conservation and utilization. FRIN has modest human and infrastructural capacities for tissue culture, molecular marker applications and genetic engineering research activities, such as genetic modification via somatic embryogenesis. The National Biotechnology Development Agency is also well equipped with molecular genetics, plant tissues culture and genetic modification facilities that are available for use in both crop and forestry research.

The only country in SSA that falls in the 'high' capacity category is South Africa. The country has human and infrastructural capacities for a wide range of biotechnology research and applications (Mulder 2003; Liebenberg 2013; Liebenberg et al. 2013). South Africa's Institute for Commercial Forestry Research (ICFR) is an independent provider of project-based research solutions and other related services in support of forest management for economic, social and environmental benefit in southern Africa. The ICFR works closely with other research institutes and universities. Indeed, most of the capacities for forestry biotechnology are in the specialized institutions of the ARC and the universities. For example, the Forestry and Agricultural Biotechnology Institute (FABI), which operates under ARC, and Forest Molecular Genetics Laboratory at the University of Pretoria have personnel and facilities to conduct low-, medium- and high-level biotechnology research and applications, including genome sequencing, genotyping using molecular markers, genomics and bioinformatics. The facilities and equipment include high-throughput DNA fingerprinting and genotyping research facilities, such as next-generation Illumina sequencers, HiSeq2500 and MiSeq systems. They provide commercial next-generation sequencing and SNP genotyping services. They also have webservers and commercial software packages, as well as a GALAXY workflow environment, to facilitate modular development tools and pipelines for high-throughput genomics analysis, gene annotations and bioinformatics (FABI 2016).

6.2.3 Networks and Networking

Within the forestry sector, international and national networks have significantly contributed to forestry research and development. At the international level, 17 SSA countries are members of the International Union of Forestry Research Organizations, a non-profit, non-governmental international network of forest scientists that promotes global cooperation in forest-related research and enhances the understanding of the ecological, economic and social aspects of forests and trees.

The African Forest Forum (AFF) is a continental association of individuals and institutions committed to sustainable management, wise use and conservation of forest and tree resources. Membership of AFF is open to individuals from academic, governmental, non-governmental, private sector and farmer organisations and any other entity concerned with the promotion of, support of or research into forests, forest management and forest products and also trees outside forests. In 2020, AFF currently had over 2000 members, with 94% from 51 African countries and 6% from overseas (covering 33 countries). AFF facilitates networking including for collaborative research among stakeholders and resource mobilization to support priority activities.

The Forestry Research Network for Sub-Saharan Africa (FORNESSA) is a non-profit, non-governmental scientific organization open to forestry and forest-related organizations and individuals. It embodies three sub-regional networks: the Association of the Forestry Research Institutions of Eastern Africa, which has ten member countries; the Forest Research Network of the Conférence de Responsables de Recherché Agronomique Africains, which has 20 member countries; and the Southern African Development Community (SADC), which represents research institutions in the 14 SADC states. The main objective of the network is to support and strengthen forestry research in order to contribute to the conservation, sustainable management and utilization of forest resources in SSA. Members include CSIR-Forestry Research Institute of Ghana (FORIG); Kenya Forestry Research Institute (KEFRI); Faculty of Forest Resource Technology, Ghana; Forestry Research Institute of Malawi; Institut de l'environnement et des Recherches Agricoles, Burkina Faso; Forestry Development Authority, Liberia; and Institut de Recherche Agricole pour le Développement, Cameroon, among others. FORNESSA has an online information service, FORNIS (https://www.fornis.net/), which serves as a platform for access and exchange of forest-related scientific information among partner organizations and facilitates the promotion of research of member institutions.

Another noteworthy network, and one which is of direct relevance to biotechnology research and applications in forestry, is the African Forest Research Network (AFORNET). This is a network of African forest-research scientists formed to support and encourage African scientists to carry out high-quality forest research. AFORNET runs a research grants scheme which aims to strengthen research capacities of individual scientists. The programme has awarded over 140 research grants in areas such as forest management, timber and fuel production, forest genetics, agroforestry, biodiversity and socio-economic studies.

The other network of direct relevance to biotechnology is the Sub-Saharan African Forest Genetic Resources (SAFORGEN) programme. SAFORGEN was established in 1998 by francophone countries with anglophone countries joining in 1999, and the network covers three sub-regions of SSA, namely, West and Central, Eastern and Southern Africa. Its purpose is to study tree genetic resources, to develop strategies and approaches for their conservation and sustainable management and to disseminate knowledge and raise awareness among relevant national and international stakeholder. It is one of four regional networks around the globe, the FORGEN networks, which were established by Bioversity International, a CGIAR centre, in collaboration with FAO and national research institutes and ministries responsible for forestry in FAO member countries, focusing on scientists and managers involved in studying and managing genetic resources of trees. SAFORGEN has a membership of 20 SSA countries and seeks to enable people and institutions in the region to undertake assessments of processes that shape forest genetic diversity at population and landscape levels; develop strategies, methods and tools for the conservation and sustainable use of forest genetic resources; and disseminate knowledge and information about conservation and sustainable use of forest genetic resources.

SAFORGEN's research activities focus on priority species groups, particularly food, medicinal, aromatic, wood, fibre and fodder tree species. SAFORGEN has been active in contributing to the development of the FAO's State of the World's Forest Genetic Resources (SoW-FGR) series of publications and is seen as a key player in the implementation of the Global Plan of Action for the Conservation, Sustainable Use and Development of Forest Genetic Resources which is a strategic framework based on the findings of the SoW-SGR.

In 2016 SAFORGEN members met in Douala, Cameroon, and identified SSA priorities among the broader strategic priorities of the global plan of action. The priorities for SSA were to develop and reinforce research programmes on tree breeding, domestication and bio-prospection in order to unlock the full potential of forest genetic resources; develop and reinforce national seed programmes to ensure the availability of genetically appropriate tree seeds in the quantities and of the (certified) quality needed for national plantation programmes; promote the establishment and the reinforcement of forest genetic resources (FGR) information systems (databases) to cover available scientific and traditional knowledge on uses, distribution, habitats, biology and genetic variation of species and species populations; establish and strengthen national FGR assessment, characterization and monitoring systems; strengthen the contribution of primary forests and protected areas to in situ conservation of FGR; promote the establishment and development of efficient and sustainable ex situ conservation systems, including in vivo collections and gene banks; update FGR conservation and management needs and integrate them into wider policies, programmes and frameworks of action at national, regional and global levels; develop national strategies for in situ and ex situ conservation of FGR and their sustainable use; and establish and strengthen educational and research capacities on FGR to ensure adequate technical support to related development programmes. Members of SAFORGEN include Benin, Burkina Faso, Chad,

Congo (Brazzaville), Ethiopia, Gambia, Ghana, Guinea (Conakry), Kenya, Madagascar, Mali, Niger, Nigeria, Senegal, South Africa, Sudan, Togo and Uganda.

In addition, CGIAR centres with a mandate on forestry have, especially over the last three decades, contributed significantly to forest research and development in SSA, working in partnership with national governments through the forestry and agriculture research and extension agencies as well as universities. The collective work of the CGIAR centres in this domain has particularly contributed to awareness about, and NARS capacities (of individual scientists and institutions) to work towards, improving natural resources and ecosystem services. In particular, the World Agroforestry Centre (ICRAF) and Bioversity International have assisted in capacity development and also supported research and development related to forest biotechnology and tree genetic resources.

With headquarters in Nairobi, Kenya, ICRAF prides itself as the only institution that does globally significant agroforestry research in and for all of the developing tropics (with significant presence in SSA), generating knowledge that enables governments, development agencies and farmers to utilize the power of trees to make farming and livelihoods more environmentally, socially and economically sustainable at scales. The focus of ICRAF's work is aimed at delivering increasing food and nutritional security and improved natural resource systems and environmental services. ICRAF's climate adaptation research, where biosciences play an important role, examines how trees reduce local temperatures, modulate water flow and continue to yield products after annual plants and livestock have been severely affected by climate change. The work also looks at transition to a circular bioeconomy through integrated production of food, bioenergy and renewable biomaterials. Other topics include soil organic carbon accounting for climate change mitigation and adaptation, soil fertility (including biological nitrogen fixation) and soil biodiversity and impacts on nutrients, pollutants, water and gas fluxes.

Although headquartered in Bogor (Indonesia), the CGIAR Centre for International Forestry Research (CIFOR) has a significant footprint in Africa with offices in Kenya and Cameroon. It works with the African forest research community to identify and address technical and policy gaps to ensure the sustainable management of forest landscapes so that biodiversity and communities can thrive and includes generating, using the tools of biosciences as appropriate, and disseminating research findings and solutions with stakeholders, such as policymakers, communities and practitioners, relevant for sustainable management of Africa's forests (CIFOR 2005).

6.3   Classification of Countries on the Basis of Capacities for Biotechnology

SSA countries were classified into the five categories for biotechnology capacity for forestry, summarized in Table 6.1. Most (23) SSA countries are in the category of 'very low' capacity for biotechnology research and applications in forestry. Another 16 SSA countries fall in the category of 'low' capacity. Three countries—Ethiopia,

Table 6.1 Classification of countries on the basis of capacities for biotechnology in forestry

Capacities category	Countries
Very low	Angola, Benin, Burkina Faso, Botswana, Burundi, Chad, Central African Republic, Republic of Congo, Djibouti, Equatorial Guinea, Eritrea, Eswatini, Gambia, Gabon, Guinea, Guinea Bissau, Lesotho, Liberia, Niger, Sierra Leone, Somalia, South Sudan, Togo
Low	Cameroon, Côte d'Ivoire, DRC, Ghana, Madagascar, Namibia, Malawi, Mali, Mozambique, Rwanda, Senegal, Sudan, Tanzania, Uganda, Zambia, Zimbabwe
Medium	Ethiopia, Kenya, Nigeria
High	South Africa
Very high	None

Countries not listed due to lack of sufficient data: Cape Verde, Comoros, Mauritania, Mauritius, Sao Tome and Principe and Seychelles

Kenya and Nigeria—are in the 'medium' capacity category. In these countries, the use of low-level biotechnology research and applications, such as bio-fertilizers and tissue culture, is quite widespread, almost routine, and there is also substantial capacity for research and application of medium-level biotechnology, especially MAS and some emerging capacities for genetic engineering, especially somatic embryogenesis (Abraham 2009; IFPRI 2014). Only South Africa falls in the 'high' capacity category, while no countries were categorised as having very high capacity for biotechnology for forestry.

6.4 Enabling Environment for Biotechnology Applications in Forestry

6.4.1 Public Awareness and Political Support

The same institutions described under crops shape the public awareness and political economy that influence enabling environment for the application of biotechnologies in forestry. However, there are forest-specific institutions, including networks (Sect. 6.2.3 on capacities) that focus on creating awareness about the importance of forests more generally and the required research and development actions. In so doing, they also contribute to shaping the overall operational environment that is relevant for biotechnology applications. Some of these are summarized below.

6.4.2 Biotechnology Policy and Biosafety Frameworks

There are no regulations in any SSA country specific to the use of GM forest trees. Although policies and regulations adopted for agricultural crops are generally extended to cover forest trees, they present special challenges because of the long timeframes and lifespans and because trees are often a wild resource and major constituents of an ecosystem. Forests are not only trees, and forest ecosystems are more fragile, longer-living and less closely controlled than crop fields (FAO 2011b). Decision-making is also complicated by the fact that, while agriculture is primarily viewed as a production system, forests are generally viewed as a natural system, important not only for the conservation of biodiversity but also for social and cultural values. The use of GM forest trees is therefore usually viewed more as a political and environmental issue than as a technical or trade issue (FABI 2016).

As pointed out under crops, the majority of SSA countries have National Biosafety Frameworks; many have policies, acts and the regulatory institutions as well. These frameworks are all encompassing and are not sector specific. Forestry biotechnology research and development is governed by the same legal and policy framework as for the crops sector. However, in some countries, sectoral policies make distinctions or exceptions. In The Gambia, for example, the Forest Policy of 2010–2019 encourages use of biotechnologies to improve the forestry sector. In Kenya, there are policies establishing a lead research institute on forest tree biotechnology: the Kenya Forest Research Institute (KEFRI) was established as a State Corporation in 1986 under the Science and Technology Act, which has since been repealed by the Science, Technology, and Innovation Act No. 28 of 2013. In many countries, while policies governing forestry research activities exist, in general they are not specific to biotechnology.

6.4.3   Public and Private Sector Investments

The limited available information reveals that the level of public and private investments in forestry in general is low and it is even lower for forestry biotechnology. Moreover, private sector involvement in forestry R&D is essentially non-existent in most SSA countries. Although no specific figures are available, the main form of public investments in forestry biotechnology—where this exists—is through the establishment of forestry research institutes, many of which are poorly resourced, have insufficient facilities and have therefore only developed low-level biotechnology activities. The only exception is South Africa which has invested substantially in forestry biotechnology through the establishment of the Forestry and Agricultural Biotechnology Institute (FABI), a full-fledged forestry biotechnology research and development institute with cutting-edge facilities and staff to undertake low-, medium- and high-level biotechnology R&D activities (FABI 2016) (Box 6.1).

Box 6.1 Creating an Enabling Environment for Agricultural Biotechnology Through an Integrated Cross-Sectoral R&D Approach: The Case of FABI

The Forestry and Agricultural Biotechnology Institute (FABI) in South Africa is an example of an integrated approach to biotechnology institutional capacity building that incorporates development of facilities, training, research services and establishment of partnerships.

Run as a post-graduate research institute, it was established in 1997 as one of the institutes of the Agricultural Research Council (ARC) based on a recognition that the future of forestry and agriculture in South Africa depends strongly on the incorporation of new technologies into these sectors.

FABI delivers its mandate through goal-directed research undertaken in partnership with major players in agricultural and forestry biotechnology. Being located at the University of Pretoria enables FABI to enjoy collaboration and linkage with academia and the majority of statutory bodies undertaking research in the plant and animal sciences.

FABI has grown rapidly and currently has a complement of more than 40 academic staff, plus over 200 post-graduate students (MSc and PhD), 35 postdoctoral fellows and visiting scientists (FABI 2021). The Institute has ten research groups, each specializing in and advancing research in different aspects of forestry and agricultural biotechnology. These include the Centre of Excellence in Tree Health Biotechnology, the Forest Molecular Genetics Programme and Molecular Plant Pathology.

FABI partners with private sector companies, such as Sappi Southern Africa and Mondi Forests, to develop capacity for the application of tree biotechnology in operational tree improvement programmes. The Institute has both human capacities and facilities for high-throughput DNA marker analysis for genetic mapping, fingerprinting and genome-wide association. It also has a new single nucleotide polymorphism (SNP) genotyping platform for commercially grown eucalypts, next-generation Illumina sequencers, a HiSeq2500 and MiSeq systems, all of which enable the Institute to conduct research and provide commercial services. The Institute also has a high capacity for bioinformatics that allow for the analysis of genome information for hundreds of crops and trees species. It has implemented web services and a wide range of software and genomic databases that are used for high-throughput downstream genomics analyses, including gene and genome annotation, gene expression analysis, detection of genomic variants and discovery of genetic associations.

The well-established plant transformation platform and requisite human capacity allows the Institute to conduct pioneering high-level forestry biotechnology research. These facilities are also used for medium-level biotechnology research, such as tissue culture and in vitro propagation for a wide range of crops and forestry species.

6.5   Classification of Countries on Basis of Enabling Environment for Biotechnology

The classification of SSA countries on the basis of enabling environment for forestry biotechnology is summarized in Table 6.2. No countries were classified as 'very strong', while Ethiopia, Kenya, Nigeria, South Africa and Sudan were 'strong'. Botswana, Ghana, Malawi, Mali, Namibia, Tanzania, Uganda and Zimbabwe were classified a 'medium', and all other countries for which information was available (31) were either 'weak' or 'very weak'.

6.6   Applications of Biotechnology in Forestry

The technologies for forestry on the basis of which countries have been assessed and classified in this study are summarized in Table 6.3.

Table 6.2 Classification of countries on the basis of enabling environment for biotechnology applications in forestry

Enabling environment	Countries
Very weak	Angola, Benin, Burundi, Chad, Central African Republic, Republic of Congo, Djibouti, DRC, Eritrea, Gambia, Equatorial Guinea, Eswatini, Gabon, Guinea, Guinea Bissau, Lesotho, Liberia, Namibia, Niger, Sierra Leone, Togo, Somalia, South Sudan
Weak	Burkina Faso, Cameroon, Côte d'Ivoire, Madagascar, Mozambique, Rwanda, Senegal, Zambia
Medium	Botswana, Ghana, Malawi, Mali, Namibia, Tanzania, Uganda, Zimbabwe
Strong	Ethiopia, Kenya, Nigeria, South Africa, Sudan
Very strong	None

Countries not listed due to lack of sufficient data: Cape Verde, Comoros, Mauritania, Mauritius, Sao Tome and Principe and Seychelles

Table 6.3 Biotechnologies on the basis of which countries have been classified for forestry

Level of technology	Low-tech	Medium-tech	High-tech
Examples of technologies	• Bio-fertilizers • Vegetative propagation • Tissue culture	• DNA fingerprinting and genotyping • High-throughput DNA marker analysis for genetic mapping	• Genetic engineering • Whole genome mapping

6.6.1 Low-tech Applications

Low-tech applications in forestry include low-cost vegetative propagation, bio-fertilizers, micro-propagation based on tissue culture and use of molecular markers for genetic fingerprinting.

Bio-fertilizers and Vegetative Propagation Low-level biotechnologies, particularly bio-fertilizers and vegetative propagation, have a long tradition of use in SSA. In eastern and southern Africa, these technologies are widely used in improving forestry management and productivity. In Kenya, for example, Kenya Forestry Research Institute (KEFRI) has developed a bio-fertilizer, marketed as Kefrifix, for enhanced growth and productivity of forest species (UNESCO 1994; Odame 1997; Omondi 2014). Other SSA countries have also been developing and applying low-level technologies in forestry over the last few decades. For example, in Botswana, the Department of Forestry and Range Resources has been using vegetative propagation methods in forestry species. In Tanzania, bio-fertilizers developed through nitrogen fixation and mycorrhizal inoculation have been used in forestry productivity. Namibia, the driest SSA country, is giving significant attention to forestry and is using low-tech applications, especially grafting and other forms of vegetative propagation of key fruit tree species.

In South Africa, low-tech applications, such as development of hybrids through research-based vegetative propagation, have grown in use over the past two decades. Private sector companies such as Sappi Southern Africa and Mondi Forests have invested in research and development as well as commercial application of vegetatively propagated hybrids of *Pinus* spp. and *Eucalyptus* spp. (Wingfield and Wingfield 2003). Bio-fertilizer application has yielded positive results for indigenous forest species in the eastern Madagascar littoral forests as well as for exotic forest species, including eucalypts, acacia and cypress. Other countries in the region that have undertaken research and applied vegetative propagation and bio-fertilizer technologies are Eritrea, Malawi and Sudan (Olembo et al. 2010).

In West and Central Africa, low-tech applications, such as bio-fertilizers and vegetative propagation, are applied widely. In Nigeria, the Forestry Research Institute routinely uses vegetative protocols for propagation of medicinal and endangered trees. In Côte d'Ivoire, forestry research organizations routinely use industrial propagation through rooted cuttings in forestry species such as rubber, cocoa and coffee. In Gabon, there are various projects on biological nitrogen fixation and production of legume inoculants for forestry medicinal plants. The Forestry Research Institute under the CSIR in Ghana undertakes substantial work in research and application of industrial and vegetative propagation through rooted cuttings in a wide range of forestry species, including rubber, cocoa, oil palm and coffee. In Mali, the Institut d'Economie Rurale (IER) has conducted a series of vegetative propagation experiments on 12 fodder tree species to identify the most appropriate multiplication methods. Moreover, in the quest to adopt tree-based food banks for food security in Mali, IER in collaboration with ICRAF has undertaken research on superior accessions and grafting of five priority tree species including baobab

(*Adansonia digitata*), jujube (*Ziziphus mauritiana*), moringa (*Moringa oleifera*), tamarind (*Tamarindus indica*) and shea (*Vitellaria paradoxa*). The technology has been tested and is being adopted by farmers.

Tissue Culture Techniques Tissue culture is a technology that is used widely in SSA. It has a fairly long tradition of use in several countries though application of the technology in forestry research is less than in the crops sector. The different techniques in use include mass propagation of elite genotypes and multiplication of threatened indigenous forestry tree species. Other methods that are widely used include disease elimination through meristem culture as well as in vitro propagation and conservation (Olembo et al. 2010; Masiga et al. 2013).

Many countries in West and Central Africa have been conducting tissue culture research in forestry species. Nigeria is among the leading countries in research and application of this technology in forestry. The leading forestry research institutions, FRIN and the Rubber Research Institute of Nigeria, have also been conducting research to develop tissue culture and propagation protocols for medicinal and endangered trees. In Côte d'Ivoire, tissue culture, particularly in vitro propagation and clonal propagation, is routinely applied in industrial forestry species by the CNRA. The technology is also used in micro-propagation of forestry species such as *Eucalyptus* and *Acacia* spp. In DRC, the University of Kisangani undertakes research work in in vitro propagation of *Acacia* and *Leucaena* species as well as medicinal plants such as *Nauclea* spp. and *Phyllanthus* spp. Other countries with forestry tissue culture activities include Cameroon, Ghana and Gabon.

Tissue culture techniques have been used even more widely in forestry research in eastern and southern Africa. Forestry research institutions and universities undertake most of the research work (Masiga et al. 2013). Kenya is among the most advanced in research and application of tissue culture in forestry as well as in crops. The Kenya Forestry Research Institute biotechnology laboratory carries out research routinely in tissue culture (micro-propagation and clonal multiplication) of elite forestry genotypes of such species as *Melia volkensii*, *Acacia* spp. and several indigenous tree species. The tissue culture technology uptake in Kenya is quite high as evidenced by emerging private enterprises such as the Tree Biotechnology Project Trust (TBPT), a private company that is developing clonally propagated tree clones of *Eucalyptus* spp. and several indigenous tree species (TBPT 2010). The uptake of tree biotechnology approach used by the TBPT has now spread to Tanzania.

The Forestry and Agricultural Biotechnology Institute (FABI) of South Africa collaborates with forestry companies, such as Sappi Southern Africa and Mondi, to develop capacity and resources for the application of biotechnology in operational tree improvement programmes. The Institute is undertaking comprehensive research and development activities on tissue culture and in vitro propagation of superior genotypes of several commercial forestry species, including *Eucalyptus* spp. and *Pinus* spp. (FABI 2016).

Substantial research and application activities are being conducted in other countries in the region as well. For instance, in Madagascar, micro-propagation techniques are being used for the development of disease-free tropical trees. In

Zambia, the Tree Improvement Research Centre at the National Institute for Scientific and Industrial Research is using tissue culture techniques for the domestication of wild fruit trees. Similarly, the National Agricultural Research Organization (NARO) in Uganda is using tissue culture for clonal propagation of commercial forest trees. In Zimbabwe, tissue culture is applied in forestry research by the Forestry Commission of Zimbabwe for clonal propagation of superior and elite genotypes of tree species as well as genetic improvement for drought and disease tolerance. Tissue culture techniques, especially micro-propagation of fruit trees and clonal propagation of industrial and commercial forestry species, are also being used at a smaller scale in other countries, including Botswana, Ethiopia, Eritrea, Namibia and Sudan.

6.6.2   Medium-tech Applications

Medium-tech applications include somatic embryogenesis (a tissue culture technique), molecular markers and QTL analyses as well as functional genomics. The application of these medium-level technologies is quite limited in SSA, largely due to inadequate human and infrastructural capacities. South Africa is in the forefront in research and application of medium-level technologies. The correct identification of clones is currently the most common application of molecular markers for genetic fingerprinting and DNA analysis in operational breeding and wood production. These applications are routinely used by FABI in collaboration with several forest companies in South Africa (Mulder 2003; FABI 2016). They include DNA fingerprinting and genotyping (microsatellite and SNPs) for identification and parentage analysis of eucalypts and pine trees; high-throughput DNA marker analysis for genetic mapping and genome-wide association to link genomic information to phenotypic variation in eucalyptus breeding populations; population genomics, genomic mapping of growth and wood quality traits in eucalyptus hybrids, genome-wide SNP marker discovery and genotyping in eucalyptus and system genetics analysis of cellulose and xylan biosynthesis in eucalyptus; functional genomics, carbohydrate active enzyme (CAZyme) genes involved in wood formation and transcriptome-wide prediction of xylem proteome variation in eucalyptus; functional genetics, functional analysis of secondary cell wall (SCW) transcriptional network in eucalyptus; and functional genetics of the SCW-related proteins of unknown function.

A few other countries are developing and applying medium-tech in the forestry sector. For example, the Forestry Research Institute of Malawi collaborates with Lilongwe University of Agriculture and Natural Resources (LUANAR) and the University of Malawi on forestry biotechnology, including research and development on genetic and molecular marker studies in various forestry species. In Sudan the Forest Research Center of the Agricultural Research Corporation (ARC) undertakes molecular and genetic diversity studies using SSRs and RAPDs in forestry and industrial forestry species, while in Uganda the National Forestry

Research Institute is using plant molecular genetics and MAS for breeding and improving clonal eucalyptus germplasm (Olembo et al. 2010; Ali 2009).

Research and application of marker technology in West and Central Africa are much more limited. In Cameroon, the Institut de Recherche Agricole pour le Développement (IRAD) conducts molecular and genetic diversity studies using SSRs and RAPDs in forestry species, while in Ghana, the Forestry Research Institute of the Council for Scientific and Industrial Research (CSIR) conducts molecular and genetic diversity studies using SSRs and RAPDs in forestry species. The Institute also conducts studies of pollen gene flow and molecular studies in forestry species. The only other reported research and application of molecular makers in the region is in Nigeria where the Forestry Research Institute and the National Agricultural Biotechnology Development Agency are conducting molecular diversity studies in indigenous trees species.

6.6.3 High-tech Applications

High-tech applications include backward and reverse genomics, whole-genome sequencing and genetic modification of forest trees (FAO 2011a). In this group of technologies, most tree species used in planted forests have been successfully modified at the experimental level, and traits that have been the subject of extensive research include stem shape, herbicide resistance, flowering characteristics, lignin content and insect and fungal resistance (FABI 2016; FAO 2011b). There are also DNA-based and biochemical markers that are available globally for several tropical species for management of naturally regenerated forests (FAO 2011a). The technology is also advancing from molecular markers to genomics. Biotechnology tools such as molecular markers and genomics are providing much needed knowledge about naturally regenerated tropical forests. They are also giving important insights into the nature of the entire tropical forest ecosystems. These insights include the relationship between forest trees and the microbial communities they interact with, which in turn could influence the strategies employed for managing tropical forests (FAO 2011c). However, application of these technologies in forestry is extremely limited in SSA, with the exception of South Africa.

In South Africa, the R&D work is mainly conducted by FABI which was established as a post-graduate research institute in the University of Pretoria in 1997. The institute, in collaboration with other public research institutes as well as private commercial forestry companies, conducts research in a wide range of high-end biotechnology applications. These include genetic engineering of forest trees, especially genetic control of wood development in fast-growing plantation trees, primarily eucalypts and tropical pines (FABI 2016). The institute also conducts research on high-end tree genomics and functional genetics, including significant contribution—together with the Genomics Research Institute (GRI) of the University of Pretoria—to the whole-genome sequencing of *Eucalyptus* genome done as part of global Eucalyptus Genome Initiative, bioinformatics for gene and genome annotation, gene expression analysis and detection of genomic variants and discovery of genetic associations. This work led to the development of high-throughput DNA marker resources (e.g. Diversity Arrays Technology (DArT) and single nucleotide polymorphism—SNP) for eucalyptus species and contributed to development of a eucalyptus SNP chip which will be of commercial importance and accessible through a large agri-genomics service provider (GeneSeek). Eucalyptus are widely cultivated due to the desired economic viable traits such as being fast-growing sources of biomass and essential oil producers, and so this work has been of significance not only to Africa but to the world. Collectively, the work has allowed for high-throughput analysis of wood physical and chemical properties to support tree breeding and genomic research in experimental tree populations. The research has made significant advances in the understanding of the biology of the most widely planted hardwood fibre crop in the world. Together with the completed genome *Populus*, this genome resource will serve as a model and reference for the study of fast-growing woody plants that are used as renewable feed stocks for a growing number of bio-based products, such as timber, pulp, paper, cellulose, textiles, pharmaceuticals and bioenergy.

FABI and other institutions also collaborate in research and application of other high-level technologies, such as genetic transformations, in a limited number of commercial forestry species such as *Pinus* spp. and *Eucalyptus* spp. (Wingfield and Wingfield 2003; FABI 2016).

Other than the examples in South Africa, most SSA countries have little or no capacities for research and application of these higher-end biotechnologies in forestry, though these capacities for genetic transformations are slowly emerging in Ghana, Kenya and Nigeria.

6.7 Classification of Countries on the Basis of Biotechnology Applications in Forestry

On the basis of the above analysis, and using the classification framework for applications, SSA countries can be classified into biotechnology categories as presented in Table 6.4. Out of the 44 countries with information, there was no

Table 6.4 Classification of countries on the basis of biotechnology applications in forestry

Applications category	Countries
Very low use	Angola, Benin, Burkina Faso, Burundi, CAR, Chad, Congo, DRC, Eswatini, Ethiopia, Equatorial Guinea, Gambia, Guinea, Guinea-Bissau, Lesotho, Liberia, Mali, Mozambique, Namibia, Niger, Rwanda, Somalia, South Sudan
Low use	Botswana, Cameroon, Djibouti, Eritrea, Gabon, Côte d'Ivoire, Malawi, Mauritius, Senegal, Tanzania, Togo, Zambia
Medium use	Ghana, Kenya, Madagascar, Nigeria, Sierra Leone, Sudan, Uganda, Zimbabwe
High use	South Africa
Very high use	None

country in the 'very high' use, only one (South Africa) in the 'high' use category, 8 in 'medium' and the rest in 'low' (12) and 'very low' (23) categories.

6.8   Biotechnology Impact Case Studies in Forestry

6.8.1   The Tree Biotechnology Project in East Africa

The Tree Biotechnology Project began in 1997 and was a partnership between the International Service for the Acquisition of Agri-biotech Applications (ISAAA), the Gatsby Charitable Foundation and Mondi Forests, a private South African company. Initially, the project involved transferring clonal technology to the forestry sector, first in Kenya and later in Tanzania and Uganda. Clonal technology ensures that all trees produced have the same genetic material which helps ensure consistent growth and other characteristics.

The project began by importing cuttings of hybrid eucalyptus species developed in South Africa by Mondi Forests. These cuttings were provided free of charge by the company as part of its corporate social responsibility activities.

The eucalyptus hybrids had been developed by Mondi Forests for characteristics such as increased productivity, high calorific value and improved drought resistance compared with non-hybrid plants. Three main hybrid clone groups were produced: grandis-camaldulensis (GC), a cross-breeding between *Eucalyptus grandis* and *Eucalyptus camaldulensis*; grandis-urophylla (GU), a cross-breeding of *Eucalyptus grandis* and *Eucalyptus urophylla*; and grandis-tereticornis (GT), a cross-breed of *Eucalyptus grandis* and *Eucalyptus tereticornis*.

The eucalyptus cuttings were used to establish clonal hedges in tree nurseries in the three East African countries. Cuttings were then harvested from the hedges, placed in rooting medium and then transferred for growing out. The project supported long-term trials in which the different hybrids were screened for suitability in different agro-ecological zones to enable the most appropriate hybrids to be chosen.

Although the same approach was used in Kenya, Tanzania and Uganda, the project was considered to be especially successful in Uganda. This was attributed partly to the quality of the clones and their suitability for growing in the climate and conditions found in Uganda. The decision to grow GU hybrids proved to be fortuitous: the GU hybrids were found to be resistant to an invasive pest, the blue gum chalcid, a type of gall wasp originally from Australia, which was first detected in Africa in the 2000s.

Other non-technology-focused factors that contributed to success in Uganda include the coincidental emergence of a favourable policy environment, which made government land available for leasing for forestry at scale, and the active involvement of Gatsby Clubs. The latter are private sector clubs funded by the Gatsby Charitable Foundation and formed by local entrepreneurs who have both capital and interest in investing in new business opportunities. In the Uganda Tree

Biotechnology Project, the clubs took on the task of disseminating clones and promoting private participation in the nursery business.

An additional factor in Uganda was the Sawlog Production Grant Scheme (SPGS). Independent of the Tree Biotechnology Project, the SPGS provided grants to cover half the cost of planting plantations of at least 25 hectares. To access SPGS grants for eucalyptus plantations, grantees had to obtain their saplings from certified clonal eucalyptus nurseries, thus creating demand.

Also, the Uganda government's rural electrification programme was initiated in the late 1990s. This created strong demand for transmission poles which previously had been imported from South Africa. The GU clonal technology proved well suited to producing high-quality, consistent eucalyptus poles which could be harvested earlier than non-clonal local varieties.

In 2019, FAO reported that there were 62 clonal nurseries operating in the country as well as many commercial integrated growers, including some large companies, producing sawlogs, transmission poles and other products. By then the nurseries were operating independently of support from Gatsby.

A major lesson from the project was the importance of sector conditions, such as the availability of land and finance, effective governance structures, dynamism of local firms and availability of targeted business support services. In the case of the Tree Biotechnology Project in Uganda, however, luck played a large part as the policy initiatives that created the enabling environment for the project to take off were outside its control and had not been anticipated as the project was planned and implemented.

Initially, the Tree Biotechnology Project was primarily targeted at poverty allevi- ation with a focus on smallholders and rural populations. The aim was to bring about improved access to affordable wood products and create wealth at the household level by promoting small-scale forestry as a cash crop. However, as the project developed and responded to emerging opportunities, such as the SPGS, it shifted towards a focus on private sector development and larger-scale commercial forestry.

A weakness of the project identified in a recent review was the reliance of the Ugandan clonal forestry industry on a narrow pool of genetic material: no new clones were imported after the initial batches provided by Mondi Forests and no research was undertaken locally to replenish and diversify. This makes the plantations planted with clonal hybrid eucalyptus vulnerable to a major pest out- break. The situation is similar in Kenya and Tanzania, and none of the three East African countries currently have the capacity to develop the new clones that are needed to make the sector more resilient in the future. An East African regional eucalyptus breeding programme has been mooted as a possible solution but has yet to find traction.

References

Abraham A (2009) Agricultural biotechnology research and development in Ethiopia. Afr J Biotechnol 8(25):7196–7204

AFF (2019) The state of forestry in Africa: opportunities and challenges. African Forest Forum (AFF), Nairobi, p 186

Ali AM (2009) Status of biotechnology in Sudan. In: Madkour M (ed) Status and options for regional GMOs detection platform. A benchmark for the region. FAO, Rome

CIFOR (2005) Contributing to Africa's development through forests strategy for engagement in sub-Saharan Africa. Center for International Forestry Research, West Bogor. Available via CIFOR. https://www.cifor.org/knowledge/publication/1773/. Accessed 30 June 2021

FABI (2016) Forestry biotechnology activities in Forestry and Agricultural Biotechnology Institute. FABI/ARC, Pretoria

FAO (2011a) Agricultural biotechnologies in developing countries: options and opportunities in crops, forestry, livestock, fisheries and agro-industry to face the challenges of food insecurity and climate change. ABDC-10 Report. FAO International Technical Conference. Guadalajara, Mexico, 1–4 March 2010. FAO, Rome

FAO (2011b) Current status and options for forestry biotechnology in developing countries. ABDC-10 Report. FAO International Technical Conference. Guadalajara, Mexico, 1–4 March 2010. FAO, Rome

FAO (2011c) State of the world's forests. FAO, Rome

FAO (2020) State of the world's forests. Report of the Food and Agriculture Organization of the United Nations, Rome

Forestry and Biotechnology Institute (FABI) (2021) Forestry and Agricultural Biotechnology Institute, Biennial report 20017/20019. Available via FABINET. https://www.fabinet.up.ac.za/webresources/biennialreports/FABI-Biennial-Report_2018_2019.pdf

IFPRI (2014) Outputs for Africa South of the Sahara. ASTI facilitated by IFPRI. IFPRI, Washington. Available via ASTI. www.asti.cgiar.org/publications/africa-south-of-the-sahara. Accessed 29 June 2021

Liebenberg F (2013) South African agricultural production, productivity and research performance in the 20th century. PhD Thesis, University of Pretoria

Liebenberg F, Beintema NM, Kirsten JF (2013) South Africa: ASTI Country Brief No. 14. International Food Policy Research Institute and the International Service for National Agricultural Research, Washington, DC

Malhi Y, Adu-Bredu S, Asare RA et al (2013) African rainforests: past, present and future. Philos Trans R Soc B 368(1625):20120312. https://doi.org/10.1098/rstb.2012.0312

Masiga CW, Mneney E, Wachira F et al (2013) Situational analysis of the current state of tissue culture application in the Eastern and Central Africa region. Association for Strengthening Agricultural Research in East and Central Africa (ASARECA), Entebbe

Mulder M (2003) National biotechnology survey. eGoliBio Life Sciences Incubator, Pretoria

Odame H (1997) Biofertilizer in Kenya: research, production and extension dilemmas. Biotechnol Dev Monit 30:20–23

Olembo N, M'mboyi F, Oyugi K et al (2010) Status of crop biotechnology in sub-Saharan Africa: a cross-country analysis. African Biotechnology Stakeholders Forum (ABSF), Nairobi

Omondi S (2014) Bio-fertilizer for enhanced growth and productivity of leguminous plants. Kenya Forestry Research Institute, Muguga

Sola P, Cerutti PO, Zhou W et al (2017) The environmental, socioeconomic, and health impacts of woodfuel value chains in Sub-Saharan Africa: a systematic map. Environ Evid 6(1):1–6. https://doi.org/10.1186/s13750-017-0082-2

Tree Biotechnology Programme Trust (TBPT) (2010) Scaling out programme of private nurseries in Kenya project: Final Technical Report. Tree Biotechnology Program Trust, Nairobi

UNDP, UNESCO (1994) African Network of Microbiological Resources Centres (MIRCENs) in biofertilizer production and use. Project findings and recommendations, UNDP, Paris

Wingfield BD, Wingfield MJ (2003) Forest biotechnology: a South African perspective. South Afr For J 199:1–5

The State of Capacities, Enabling Environment, Applications and Impacts of Biotechnology in the Sub-Saharan Africa Aquaculture Sector

J. E. O. Rege and Joel W. Ochieng

Abstract

Sub-Saharan Africa countries were categorized with regard to their capacity, enabling environment and research and applications of biotechnology in aquaculture. Capacity assessment examined human capacities, institutions and facilities, operational budgets and existence of facilitating networks. Enabling environment analysis covered public awareness, participation and acceptance, existence of national biosafety frameworks, public policy, political goodwill and public and private sector investments. Research and application of biotechnology were considered based on both the extent and context of use ranging from isolated pieces of research to routine field applications. No SSA country had capacity that could be classified as 'high' or 'very high'. Four countries—Kenya, Nigeria, South Africa and Tanzania—had 'medium' capacities. All countries with data had 'weak' or 'very weak' enabling environment for biotechnology applications in the sector, except the Republic of Congo, Kenya, Malawi, Mozambique, Namibia, South Africa, Eswatini, Uganda and Zambia all of which fell in 'medium' category. In terms of applications of biotechnology, out of the 44 countries with information, 38 were in the 'very low' or 'low' use categories and six countries (Kenya, Madagascar, Nigeria, Sierra Leone, South Africa and Tanzania) in the 'medium' category. None were in the 'high' or 'very high' categories.

J. E. O. Rege (✉)
Emerge Centre for Innovations-Africa, Nairobi, Kenya
e-mail: ed.rege@emerge-africa.org

J. W. Ochieng
Agricultural Biotechnology Programme, University of Nairobi, Nairobi, Kenya

7.1   Sub-Saharan Africa Aquaculture Sector

Fish and fish products are some of the healthiest foods, providing essential proteins, omega-3 which enhances the immune system, minerals and trace elements. They are also recognized as some of the least impactful on the natural environment. For these reasons they are vital for national, regional and global food security and nutrition strategies and have a big part to play in transforming food systems and eliminating hunger and malnutrition.

For capture fisheries, these benefits are threatened by a raft of factors that include habitat loss, invasive species and overexploitation of natural fish stocks, which are approaching the limits of their natural productivity. Currently, fish and other aquatic resources in most sub-Saharan Africa (SSA) lakes and rivers have recorded a drastic decline in abundance, with many species now either threatened or endangered (Polidoro et al. 2016). Because of this decline some communities that initially depended on capture fisheries are increasingly embracing aquaculture.

Globally, the fisheries and aquaculture sector has grown significantly in the past decades, and total production, trade and consumption reached an all-time record in 2018 (FAO 2020). However, since the early 1990s, most growth in production from the sector has been from aquaculture (114.5 million tons in 2018), while capture fisheries production has been relatively stable (at 96.4 million tons in 2018), with some growth in inland capture. Between 1990 and 2018, global capture fisheries production grew by 14%: there was a 122% increase in total food fish consumption, principally driven by a more than sixfold increase in aquaculture production during this period. Total global fish production is expected to increase from 179 million tons in 2018 to 204 million tons in 2030. World production of farmed aquatic animals has been dominated by Asia, with an 89% share in the last two decades or so. Among major producing countries, China, India, Indonesia, Vietnam, Bangladesh, Egypt, Norway and Chile have consolidated their share in regional or world production to varying degrees (FAO 2020).

The motivation for introduction of aquaculture in SSA, towards the turn of the twentieth century, was not food security but as a recreational resource for settlers (Hecht et al. 2006), for example, the introduction of American largemouth bass in Kenya and South Africa. Later introductions, however, were motivated by the need to provide food and nutrition security and incomes and to mitigate climate risks (Brummett et al. 2008). Today, Africa contributes about 3% to global aquaculture production (Halwart 2020). At the regional level, aquaculture accounts for 16–18% of total fish production in Africa (FAO 2020), which is a similar proportion to the Americas and Europe. In 2011, fisheries and aquaculture directly contributed USD 24 billion to the African economy, representing 1.3% of total African GDP. In 2018, there were about 1.2 million aquafarmers across the continent (including North Africa): an increase from 920,000 in 2014. Huge production increases have recently been reported, from 110,200 tons in 1995 to 2,196,000 tons in 2018 (Halwart 2020), partly accounted for by huge investments in Ghana, Nigeria and Uganda (Cai et al. 2017). Tilapia and catfish are the two most important farmed fish species in SSA.

With predictions of a global (Delgado et al. 2003), regional and national shortage of fish and fish products and rising prices, aquaculture will need to be developed rapidly. Under these constraints and urgencies, innovative technologies, including biotechnology, are required to shift aquaculture facilities to achieve higher productivity, profitability and resource use efficiency while enhancing ecosystem sustainability.

Aquaculture in the SSA is primarily small-scale rural. It is estimated that about 95% of total production is from small ponds, 100–500 m^2, which are often just a small part of other agricultural activities. Average yield is estimated at about 500 kg per hectare per year, although performance varies considerably from less than 100 to more than 10,000 kg per hectare per year (Machena and Moehl 2001). While there are commercial farms using cage, raceway and recirculating systems, ponds are the principal production unit.

Use of biotechnology in aquaculture has the potential to address the challenges the sector faces by enhancing production efficiency and minimizing losses. Specifically, biotechnologies now exist that can be deployed to reduce production costs by increasing growth, improving feed conversion efficiency and reducing mortalities; increase food safety; and reduce post-harvest losses.

7.2   Capacities for Biotechnology Applications in Aquaculture

Despite strong potential and significant expansion of its existing aquaculture industry, Africa accounts for only 2% of global aquaculture production. While there are many interrelated factors responsible for this situation, the two most important ones are financial investments and expertise.

The management interventions, infrastructure and supporting technologies utilized in aquaculture include a range of activities, such as seed supply and stocking, handling, feeding, controlling, monitoring, sorting, treating, harvesting, processing and use of prophylactic measures. Specific technical areas where biotechnology could make a major difference include feed, health and genetics/breeding, the latter being a critical underpinning for the production of appropriate fish seed. With regard to feed, issues include nutrient requirements of the fish in question and availability and supply of affordable fish feeds that are appropriate for the specific local farming conditions. Disease is also an important constraint to the growth of many aquaculture species and is responsible for severely impeding both economic and socio-economic development of aquaculture in many countries of the world, including SSA.

The current area under aquaculture production in SSA is estimated at between 50,000 and 100,000 hectares compared to the estimated available surface area of 30 million hectares suitable for aquaculture and an additional 12 million hectares of floodplains which is also suitable for fish production. There is also a potential for cage culture, given the availability of water bodies such as large lakes throughout the SSA region. Integration of livestock and aquaculture is practiced in many SSA countries. However, the use of livestock or industrial effluents (e.g. sewage, heated water or process water) for aquaculture will increasingly raise issues of public health, such as disease transmission and accumulation of toxic compounds. Access to supplies of suitable water (coastal, estuarine and particularly freshwater) will become increasingly challenging and is likely to be a source of widespread competition and conflict. Areas that increasingly need addressing include use of water more efficiently; integration of aquatic production with both crop and livestock agriculture; linking/sharing or complementing other water resource uses; and using water for which there is less competition due to its reduced suitability for drinking, irrigation or other agriculture uses. Generally, aquaculture requires high levels of skill and professionalism that reflects and addresses the fact that it is highly integrated with other users and because of its complex interactions with the local environment.

Most of the technologies developed for use in aquaculture, even low-tech examples, have only been applied to limited extents in SSA. This has been attributed to five interrelated factors (FAO 2015):

- Limited personnel and infrastructure capacity of national institutions to develop, adapt or use the requisite technologies.
- The extension services have limited capacity to disseminate and scale available technologies.
- Farmers in the region lack technical capacity to apply the technologies.
- Lack of an enabling policy environment to facilitate adoption and domestication of these technologies.
- Farmers are inherently unwilling to adopt the technologies and also unable to do so due to lack of finances.

7.2.1   Human Resources

The state of human resources in fisheries and aquaculture in the SSA region is summarized in Fig. 7.1.

In terms of ranking based on absolute staff numbers (and excluding South Africa which did not have comparable data), Tanzania (with total of 117 full-time equivalents (FTEs) for the sector), Nigeria (77), Kenya (69), Niger (66) and Ethiopia (60) are the top five; they are followed by Mozambique, Mauritius, Ghana, Côte d'Ivoire and DRC. Some countries prioritize this sector. For example, for Tanzania the FTE allocation represents 16% of the country's total agricultural R&D FTE. Countries allocating 10% or higher of their agricultural R&D FTEs to fisheries include Mauritius (36%), Mozambique (21%), Niger (19%), Gabon (16%) and Senegal (14%). Côte d'Ivoire, Gambia and Madagascar are each at 10%. Corresponding figures for Ethiopia, Nigeria and Kenya are 3%, 8% and 8%, respectively.

Interestingly, certain countries primarily associated with crop and livestock agriculture do recognize and invest in fisheries and aquaculture agricultural R&D as indicated by proportionate allocation of FTEs. So, Côte d'Ivoire (11 vs. 10%) and

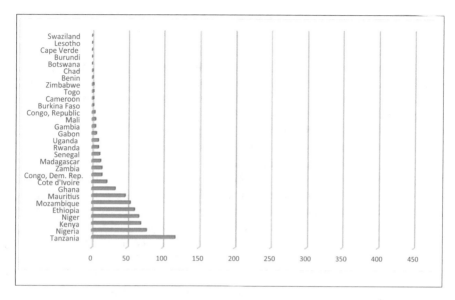

Fig. 7.1 Total fisheries and aquaculture FTEs in SSA countries (2014)

Senegal (13 vs.14%) allocate about equal FTEs to livestock as they do to fisheries and aquaculture. Figures for Mozambique are 26% for livestock and 21% for fisheries/aquaculture. Niger is 25% and 19%, respectively. Tanzania (16 vs. 11%) and Mauritius (36 vs. 8%) allocate more FTEs to fisheries/aquaculture than to livestock.

Availability and capacity of training institutions are important defining factors for biotechnology capacity. In Madagascar, for example, there are limited number of training institutions and few laboratories and field stations with capacity to conduct research. The country has a small number of national researchers, and the training levels are generally low. There has also been a decrease in public agricultural research and development budget. The National Center for Applied Research and Rural Development (FOFIFA) is the largest research institution with a broad mandate covering crop, livestock, forestry, postharvest and socio-economic research.

Whereas some of the countries have adequate universities and other training institutions specialized in fisheries and aquaculture (such as Benin, Kenya, Mozambique, Namibia, RSA, South Sudan, Tanzania and Uganda), others (such as Chad, Djibouti, Gambia and Somalia) have either no university at all with fisheries programmes or universities not specialized in any area of agricultural biotechnology (such as Burundi). Number of staff (FTEs) alone does not necessarily translate into applications. The broader institutional capacity and facilities as well as the extent to which the relevant institutions collaborate with other similar as well as stronger institutions or initiatives are a critical factor in defining overall capacity for biotech applications in a particular sector of a country. These are discussed below.

7.2.2   Institutions and Facilities

Most countries with moderate capacity to conduct research on high-tech applications are those that either host CGIAR or other intergovernmental centres or work in collaboration with universities in the North or South. This is true for crops and livestock but is also true for other sectors, including fisheries. This category of countries includes Ethiopia, Kenya, Nigeria and South Africa. Development partners (FAO, USAID, CIDA and others) and significant private sector investment in the sector also raise the research capacity of countries.

In South Africa the Agricultural Research Council aims to enhance aquaculture through research, development and technology transfer by building a sustainable aquaculture community in partnership with all stakeholders including producers, processors, communities, universities and public agencies. Improvement of nutrition and feeding is one of the major research areas. Development of feeds and feeding management practices for cultivated aquatic organisms and investigation into feed ingredients as alternatives to fish-based ingredients are the two main strands of nutrition research. Breeding and genetics are another major area of research, with a focus on genetic improvement of tilapia being done in collaboration with Chinese scientist. The initiative includes development of a selective breeding research facility which aims to facilitate the supply of quality seed to farmers and for restocking natural systems. Other areas of research include the development of backyard aquaponics systems; development of tank culture of freshwater crayfish, *Cherax cainii* (marron); and testing high-value species in aquaponics, including for essential oil and medicinal plants.

Zini Fish Farms in South Africa is 45 hectares in extent and comprises 52 half-hectare earthen ponds. This outstanding facility is situated alongside the Mlalazi Estuary in the coastal village of Mtunzini in northern KwaZulu Natal. The primary product is the saltwater tilapia (*Oreochromis mossambicus*)—popularly known as Mozambique tilapia or Blue Kurper in South Africa. The warm climate and access to sea and estuary water are also well-suited to rare and prized fish like dusky kob, spotted grunter, natal stumpnose, Malabar rockcod and mud crab. Facilities include a hatchery, nursery, workshops, cold room, feed store, administrative offices and managerial and staff accommodation. The hatchery has more than 2000 actively spawning tilapia breeders. The nursery has the capacity to restock all ponds on an annual basis and also to sell fingerlings to emerging farmers in the country. The farm actively collaborates with other aquaculture industry organizations, including local and international research institutions. Its extensive pond system and on-site accommodation offer opportunities for long-term research projects and training visits by groups of aquaculture students from tertiary academic institutions.

In Nigeria, the national competent authority responsible for the management of fisheries and for the preparation of policies and programmes for the development of fisheries is the Federal Department of Fisheries of the Federal Ministry of Agriculture and Natural Resources, which also provides technical support to State Departments of Fisheries (SDF). The latter also provides support to local government authorities on fisheries matters. Fisheries and aquaculture research is carried out by the Nigerian Institute for Oceanography and Marine Research (NIOMR) and by the National Institute for Freshwater Fisheries Research, while aquaculture training is ensured by the African Regional Aquaculture Centre (ARAC) of NIOMR. NIOMR's scientific research activities encompass ocean and marine sciences, namely, aquaculture, biological oceanography, biotechnology, fisheries resources, fish technology and product development, physical and chemical oceanography and marine geology/geophysics. The story of the development of catfish farming in Nigeria is particularly instructive and a demonstration of what national commitment can achieve (Box 7.1).

Box 7.1 How Nigeria Became the World's Largest Producer of African Catfish

The history of catfish farming in Nigeria dates back to 1951 when the British colonial government tested the feasibility of catfish farming in Nigeria at a small experimental station in Onikan, Plateau State—the Lagos and Panyam Catfish Farm. The aquaculture program was continued by the Nigerian government after independence. Experimentation at various farms continued until 1971 when demonstration farms were established in Ibadan, Akure, Umuna Okigwe, Itu and Opobo. Four zonal fingerlings production and training centres were established in Oyo, Umuna Okigwe, Panyam and Mando-Kaduna between 1978 and 1980 with the support of the United Nations Development Programme (UNDP). The African Regional Aquaculture Centre (ARAC) was established at Aluu, Port Harcourt with the assistance of FAO to provide research support and training in Nigeria and other SSA countries.

Catfish farming was initially dominated by the government, which was directly involved in experimentation, demonstration and production. The River Basin Development Authorities (RBDA) were established to run commercial catfish farms and prove the commercial viability of aquaculture. Technical assistance was also provided to support the construction of ponds and provision of inputs, including fingerlings and feeds.

During the period 1981 to 1991, there were efforts by the federal government to stimulate broader interest in commercial aquaculture. These started to yield dividends as many private investors and state governments went into catfish farming. It also became apparent that the government's role should focus on providing an enabling environment leaving production to the private sector.

Catfish farming has given Nigeria a niche in global aquaculture production. Today, Nigeria is the world's largest producer of farmed North African catfish, *Clarias gariepinus*, with production for 2020/2021 estimated to be over 250,000 metric tonnes. With catfish accounting for 80% of its aquaculture production, Nigeria is also currently the second highest producer of aquaculture products in Africa behind Egypt.

(continued)

Box 7.1 (continued)

Although Nigeria exports smoked catfish worldwide, its major export destinations are Ghana, Cameroon, Senegal and Cote d'Ivoire. Some of the latter also re-export to Europe suggesting that Nigeria could provide greater value addition from its aquaculture exports.

Source: Uwagbale (2021)

Aquaculture has grown rapidly in Kenya over the last 15 years and plays an increasingly important role in national fish supply. Freshwater fish accounts for close to 98% of Kenya's aquaculture production, and the country is ranked the fourth largest producer of aquaculture products in Africa. Production from aquaculture systems grew from 4218 tonnes in 2006 to over 24,096 tonnes in 2014, representing 15% of total national fish production. The rapid growth is mainly associated with the intervention of the government through the intersectoral Economic Stimulus Programme (ESP) in 2009, whereby KSh 22 billion (approximately USD 283 million) was channelled into key sectors from 2009 to 2012. Kenya, however, has far greater capacity for fish farming, with over 1.14 million hectares potentially available to enable a production capacity of over 11 million tonnes per year (Nyandat and Owiti 2013). Aquaculture will become increasingly important as wild fish catches continue to decline, the population grows and demand for raw material for value addition continues. The country has fast-growing fish species, such as Nile tilapia and African catfish, and extensive freshwater resources suitable for the cage, pond and tank-based aquaculture systems, and its agriculture and fisheries sectors produce most of the raw materials needed for locally made fish feeds. Local and regional market potential is also strong. There has been a steady evolution and growth of fisheries institutions in Kenya. The enactment of the Fisheries Management and Development Act in 2016 led to the establishment of the Kenya Fisheries Service, a state corporation, backed by other institutions such as the Kenya Fisheries Council and the Kenya Fish Marketing Authority. Despite sensitization on the need to have quality brooders, many farmers still use genetically inferior fish stocks or undomesticated. Availability of quality fish seed is therefore a major bottleneck to sustainable aquaculture. In 2009 Kenya had only a total of 21 hatcheries producing African catfish and Nile tilapia fingerlings. The number had increased to 147 by 2012 but dropped to 127 by 2015. The Kenyan government has designated several facilities across the country to serve as aquaculture research centres, training facilities and sources of fingerlings and feed for fish farmers. They include the National Aquaculture Research Development and Training Centre (NARDTC) in Sagana, Kisii fish farm training centre, Kiganjo trout farm, Ndaragua trout farm, Chwele fish farm, Lake Basin Development Authority in Kisumu, Wakhungu fish farm in Busia, Sangoro research station, Kegati research station and Kabonyo and Ngomeni fish farms. However, most of these centres lack laboratory equipment and human capacities to spur significant aquaculture development in their assigned domains (Munguti et al. 2014). Application of sex reversal and super YY technology in 87 of the centres remains suboptimal due to limited knowledge of the technology among the practitioners (Mirera et al. 2021). Feed remains one of the most pressing challenges to growth of aquaculture in the country.

The principal responsibility for fisheries research in Ethiopia lies with the Ethiopian Institute for Agriculture Research (EIAR). Fisheries research is also undertaken at the Aquatic Biology Unit, Department of Biology, Faculty of Science of Addis Ababa University (AAU) as well as in Alemaya Agricultural University and Debub University. The AAU's Aquatic Biology Unit has established the Freshwater Fisheries Limnology Project (FFLP) covering Lakes Awassa, Ziway, Langano and Shala. In collaboration with the Wageningen Agricultural University of the Netherlands, the AAU has been implementing a research project on Lakes Tana and Chamo and River Baro. In addition, the Research Centre at the Water Technology Institute at Arba Minch undertakes limnology studies, that is, study of inland aquatic ecosystems, of Lakes Chamo and Abaya, and Sebeta Fish Culture Centre offers training in aquaculture.

In aquaculture biotechnology, capacity for training tends to correlate positively with a country's capacity to conduct research. The capacity of public research institutions and laboratories to conduct biotechnology research in aquaculture is categorized as 'very low' or 'low' for most countries. Currently, individual researchers and research groups in national centres and universities are limited in ability to conduct high impact research and train students due to minimal laboratory infrastructure and equipment, and research funding; consequently, there were very limited number of funded research projects.

In Mozambique, for example, human and infrastructural capacity for agricultural biotech is very weak, with less than 100 experts shared across the 4 sectors (crops, livestock, forestry and aquaculture), consistent with a low number of training institutions and a weak research infrastructure. The National Institute of Fisheries Research (IIP) of Mozambique is a public institution, under the Ministry of the Sea, Inland Waters and Fisheries, that undertakes research on the fishing resources of Mozambican jurisdictional waters with a view to their management, conservation and optimization of their exploitation. Aquaculture development in Mozambique has been unstable: production was 1222 tonnes in 2005, decreased to 490 tonnes in 2009 and then recovered to reach more than 1800 tonnes in 2017. This fluctuation was caused mostly by changes in production at large-scale marine shrimps farming: in 2011, shrimp farming was devastated by a white spot disease outbreak that completely destroyed the stock, leaving shrimp farms and farm-owned processing workshops closed and workers dismissed. It is estimated that there are over 3500 freshwater fish ponds (200–400 m in area, 105 hectares) in Manica, Niassa, Tete, Sofala and Zambézia. There are currently three commercial shrimp aquaculture enterprises operating in Beira, Sofala Province (Sol and Mar with 500 ha); Quelimane, Zambézia province (Aquapesca with 1000 hectares); and Pemba, in Cabo Delgado province (Indian Ocean Aquaculture with 980 hectares). All of these use a semi-intensive farming system in earthen ponds (size range from 5 to 10 hectares) and import feed from South Africa and Seychelles or from Asia. Current production is at 4.8 tonnes/ha/year. The species produced are *Penaeus monodon*, the giant tiger prawn, and *Penaeus indicus*, the Indian white prawn.

Seaweed (*Eucheuma spinosum* and *Kappaphycus alvarezii*) is farmed in Cabo Delgado (from Pemba to Macomia, including some islands in the Quirimba archipelago) and in Nampula (between Angoche and Nacala) provinces. The fish and crustaceans species most cultivated are the giant tiger prawn (*Penaeus monodon*), Indian white prawn (*Penaeus indicus*) and the native freshwater fish Mozambique tilapia (*Oreochromis mossambicus*).

A laboratory at the Agricultural Research Institute of Mozambique (IIAM) in Mozambique was refurbished in 2004 by USAID but has limited staff and equipment; hence its utility is very low. There is need for a stronger training base in Mozambique to equip the fisheries sector with research and training capacity. IIAM holds a broad mandate focusing on crops, livestock, forestry and natural resources research. There is need for specialized centres that focus on each sector to better promote its development. Although there is a need for specialized centres for almost all countries, Mozambique, like its counterpart countries falling in the 'low' capacity category, needs to strengthen its collaboration base to improve the expertise, research and training. There is currently little evidence of robust agricultural extension and private sector involvement in fisheries and aquaculture.

7.2.3   Networks and Networking

Here networking includes collaborations, societies and consortia, which work either within a country or among countries—encompassing networks between and among researchers; among researchers, government extension and farmers; those among institutions within a country; and between a country and either North or South partners. The level of collaborations and networks in aquaculture is overall weak.

Hosting of CGIAR Centers and working with other intergovernmental institutions, such as FAO and USAID, enhanced the capacity of countries, as these centres and institutions have a focus on strengthening of research and adoption networks. In Tanzania, for example, where more than 20,000 freshwater fishponds have been constructed across the mainland, and also a large rainbow trout farm in Arusha, research institutions collaborate with WorldFish and Sustainable Aquaculture Research Networks for sub-Saharan Africa (Rukanda 2018).

The recently developed Africa Blue Economy Strategy—developed in 2019 following the Sustainable Blue Economy Conference held in Nairobi in 2018—aims, among other things, to guide sustainable utilization of aquatic resources in Africa, with a thematic area that focuses on fisheries, aquaculture, conservation and sustainable aquatic ecosystems. An EU-funded project known as 'Fisheries Governance' being implemented by African Union—Inter-African Bureau for Animal Resources (AU-IBAR) with the objective of enhancing the contribution of sustainable fisheries and aquaculture to the achievement of the AU Agenda 2063's objectives is one of the mechanisms for operationalizing aspects of the Blue Economy Strategy. Continental and regional strategies and projects such as these provide a framework for cross-country information sharing and learning, especially useful for mentoring weaker countries.

The Aquaculture Association of Southern Africa (AASA) was established in the late 1980s in order to represent the interest of the then fledgling aquaculture industry in southern Africa. The Association has since developed into a structure with representation from the various sectors contributing towards the aquaculture industry of the region, including marine species such as oysters, mussels, abalone and prawns; freshwater species such as trout, catfish, tilapia and ornamental fishes; as well as service providers such as feed companies, equipment suppliers and veterinary services. AASA's objective is to contribute towards the development of aquaculture in southern Africa through effective representation and dissemination of information.

Recognizing the potential for aquaculture to help deliver food security, create jobs and boost local business prospects throughout Africa, researchers from across the continent have recently joined forces to form the Research Network for Sustainable Marine Aquaculture in Africa (AfriMAQUA). Launched in September 2019, AfriMAQUA aims to bring marine aquaculture researchers from Southern, East and West Africa, as well as France, together to exchange knowledge, pool research efforts and strengthen capacities. The overarching goal is to facilitate scientific cooperation for the development of sustainable marine aquaculture. AfriMAQUA's launch meeting was hosted by the University of Cape Town and was attended by delegates from Cote d'Ivoire, Senegal, Tanzania, Kenya, Mauritius, Namibia and South Africa. The network is funded by The French National Research Institute for Sustainable Development (IRD). As part of larger AfriMAQUA activities, the University of Cape Town is implementing a One Health marine aquaculture project aimed at ensuring healthy production of fish from aquaculture production. Objectives include development of vaccines against emerging diseases of indigenous cultured fish, development of feed supplements for cultured abalone and sea urchins, determining the effect of seaweed and probiotic supplements on rainbow trout reared in seawater and understanding the community/population structure of microbes living on sea lettuce (of the genus *Ulva*) grown in different conditions.

7.3   Classification of Countries on the Basis of Capacities for Biotechnology

When all dimensions of capacity are taken together, most countries in SSA have either 'low' or 'very low' capacity to develop and apply or adopt, adapt and scale biotechnologies in aquaculture to reduce production costs by increasing growth, improving feed conversion efficiency, reducing mortalities, improving food safety and reducing post-harvest losses. Only Kenya, Nigeria, South Africa and Tanzania are categorized as 'medium' capacity, while no countries were categorized as 'high' or 'very high' (Table 7.1).

Table 7.1 Classification of countries on the basis of capacities for biotechnology applications in aquaculture

Capacities category	Countries
Very low	Angola, Benin, Burundi, Chad, Central African Republic, Republic of Congo, Djibouti, DRC, Gambia, Equatorial Guinea, Eritrea, Eswatini, Gabon, Guinea, Guinea Bissau, Lesotho, Liberia, Niger, Sierra Leone, Togo, Somalia, South Sudan
Low	Botswana, Burkina Faso, Cameroon, Côte d'Ivoire, Ethiopia, Ghana, Madagascar, Namibia, Malawi, Mali, Mozambique, Rwanda, Senegal, Sudan, Uganda, Zambia, Zimbabwe
Medium	Kenya, Nigeria, South Africa, Tanzania
High	None
Very high	None

Countries not listed due to lack of sufficient data: Cape Verde, Comoros, Mauritania, Mauritius, Sao Tome and Principe and Seychelles

7.4 Enabling Environment for Biotechnology Applications in Aquaculture

7.4.1 Public Awareness and Political Support

There is no comprehensive study that has analysed the level of public awareness on the existence and potential of biotechnologies to boost aquaculture production and health in SSA. Some analyses have concentrated on only one aspect of biotechnology—GM foods. Circumstantial evidence suggests that the level of awareness on biotechnology is low. A study in Nigeria, for example, found that there was little awareness about the potential of genetics in boosting fish production, and farmers are not even aware of new innovations in fish farming, thereby limiting application in the country (Erondu et al. 2011). This is consistent with the hostility with which genetically improved fish has been received in parts of SSA, including Nigeria itself, where it has been reported that many people erroneously believe that genetically improved fish will cause cancer (Erondu et al. 2011). Such scenarios have contributed towards limitation of genetic application across the country.

As has been alluded to in the earlier chapters, political support for biotechnology remains generally low in Africa—across sectors. Political support for anything, including biotechnology application, is difficult to gauge, except when inferred from intentionally designed processes. The low, albeit varied, political support for biotechnology application in SSA is generally not sector-based. A country either supports biotechnology application or not, regardless of the sector. Thus, the presence of a policy and law on biotechnology and biosafety can be interpreted as evidence of political support, except cases where these laws are enacted to prohibit the use of biotechnology in a specific domain.

7.4.2   Biotechnology Policy and Biosafety Frameworks

More than 80% of the countries in SSA have a biotechnology policy, and at least 50% have a biotechnology policy specific to the aquaculture sector. However, public awareness on these policies was moderate at best. In Mozambique, for example, there is a policy specific to aquaculture—Resolution No. 11/96, together with its implementing strategies—although even people active in the sector were unaware of this.

Although more than 90% of SSA countries have laws on fisheries, in many cases the scope and focus are limited to capture fisheries: only a few countries have laws specifically addressing aquaculture. This fact, and the content focus of the existing laws, suggests that the motivation in most countries, at least with respect to laws and regulations, is currently more concerned about conservation of natural fish sources than corresponding issues germane to the development of aquaculture. There are also instances where a country has laws but lacks policies on the sector, for example, Benin, Burkina Faso, Central African Republic, Djibouti, Liberia and Somalia. Although it would be expected that policies precede laws, the existence of a law in the absence of the relevant policy would suggest a situation where the law relied on a related policy that may not be specific to the sector, or a law whose enactment was influenced by external factors.

Countries that have explicit and specific regulations for aquaculture include Ethiopia, Madagascar and Namibia. Mali has a policy on aquaculture but does not have specific regulations for the sub-sector.

Many of the regulations that do exist around fisheries do not address biotechnology directly, but include such management aspects that ordinarily include biotechnology applications. More than 50% of the countries had their fisheries and aquaculture managed and regulated by the ministries in charge of either agriculture, livestock or fisheries or equivalent. Very few countries had agencies set up independent of these ministries to regulate the aquaculture sub-sector. This may indicate that, for most countries in SSA, the mindset is still on capture fisheries. This may explain why only four countries (Gabon, Kenya, Eswatini and Uganda) have set up institutions to specifically promote the aquaculture sector.

7.4.3   Public and Private Sector Investment

As pointed out under crops and livestock, the private sector plays important, although variable, roles in financing agricultural investments in SSA, a region whose governments face resource constraints. However, there are no publications on or readily accessible databases with comprehensive data on private sector investment in the fisheries and aquaculture sector.

Staffing levels for fisheries and aquaculture obtained from ASTI data (Sect. 7.2) indicate that the sector has low levels of staffing in the majority of countries. Although FTEs (which is only one of the aspects of investment) cannot be used to fairly interpret the totality of investments, the number is a good pointer to the low level of public investment. For example, the only countries allocating 10% or more of their total agricultural R&D budget to fisheries/aquaculture are Mauritius (36%), Mozambique, (21%), Niger (19%), Gabon (16%), Senegal (14%), Côte d'Ivoire (10%), Gambia (10%) and Madagascar (10%). SSA governments must therefore significantly increase public and private investments in biotech R&D for aquaculture. Some of the ways in which private sector can be attracted include working to achieve a higher investment protection index and legal rights index and simplifying procedures for starting relevant business in the sector and registering property.

7.5   Classification of Countries on the Basis of Enabling Environment for Biotechnology

While a few countries have commendable development and regulation framework, such as Madagascar, others have challenges that will have to be overcome to catapult the sector. Madagascar is a success story in fisheries and aquaculture, with regulations related to fishery research (shrimps) and a draft biosafety bill. Madagascar has regulation/procedures governing the use of hormones for sex inversion in aquatic species and has set up a technical committee through Interdepartmental Order No. 22914/2004 of 29/11/04 (FOFIFA Technical Committee) to promote monosex male tilapia and to regulate the sector (FAO no date). However, a liberal interpretation may see some of the policies and laws as risk-centred, focusing on risk and less attention to development. Likewise, conservation policies tend to adopt an approach that ignores environmental changes and emerging challenges and treats conservation units without regard to possible influence of environmental changes. The country has a rich biodiversity and could protect this while at the same time delivering livelihood alternatives.

Despite hopeful examples, such as Madagascar, many of the countries have specific situations that may adversely affect growth of the sector. In Uganda, for example, the passing of the Biosafety Bill (2012) only occurred in October 2017 following years of controversy. The law in current state, however, limits environmental release of GM products (animals, fish or crops). Although currently there are no GM fish in SSA, such a law may scuttle research applying high-level biotechnologies, such as genetic engineering. So, SSA countries need a legal framework, policies and action plans which also address aquaculture more specifically, and to have funding that target fisheries and aquaculture.

Thus, while some countries such as Madagascar have clearly paid attention to the policy and legislation side, other enablers that can enhance research and applications are still relatively absent.

Table 7.2 presents the classification of countries by enabling environment categories for biotech in aquaculture. No countries are classified as either 'very strong' or 'strong', while nine countries (Republic of Congo, Kenya, Malawi, Mozambique, Namibia, South Africa, Eswatini, Uganda, Zambia) are 'medium'. All other countries for which information was available (34) are classified as 'weak' or 'very weak'.

Table 7.2 Classification of countries on the basis of enabling environment for biotechnology applications in aquaculture

Enabling environment	Countries
Very weak	Angola, Benin, Burkina Faso, Chad, DRC, Eritrea, Equatorial Guinea, Guinea, Guinea Bissau, Lesotho, Niger, Togo
Weak	Botswana, Burundi, Cameroon, Central African Republic, Côte d'Ivoire, Djibouti, Ethiopia, Gabon, Gambia, Ghana, Liberia, Madagascar, Mali, Nigeria, Rwanda, Sierra Leone, Senegal, Somalia, South Sudan, Sudan, Tanzania, Zimbabwe
Medium	Republic of Congo, Eswatini, Kenya, Malawi, Mozambique, Namibia, South Africa, Uganda, Zambia
Strong	None
Very strong	None

Countries not listed due to lack of sufficient data: Cape Verde, Comoros, Mauritania, Mauritius, Sao Tome and Principe and Seychelles

Table 7.3 Biotechnologies on the basis of which countries have been classified for aquaculture

Level of technology	Low-tech	Medium-tech	High-tech
Examples of technologies	• Polyploidy • Bioremediation • Probiotics	• Marker-assisted selection • PCR • Sex reversal • Bioinformatics	• Genetic modification • Gene editing techniques

7.6   Application of Biotechnology in Aquaculture

The technologies for aquaculture on the basis of which countries have been assessed and classified in this study are summarized in Table 7.3.

Faster growth and higher resistance to disease, more usable muscle meat and greater temperature tolerance are some of the improvements that aquaculture sector seeks to achieve. Specifically, biotechnology applications in aquaculture and fisheries could address such major challenges as fish diseases, water pollution, provision of quality seedstock, and feed and fertilizers used in fishponds. Through addressing these challenges—from effective breeding methods that enable rapid genetic advances, through the development of alternative feeds, to highly successful diagnostic and therapeutic methods that facilitate prevention and control of diseases—biotechnology applications can help increase fish production while at the same time reducing the environmental and economic costs of aquaculture. Globally, achievements have been made in the recent past in increasing production of fish, through genetic and other biotechnological tools—transgenic fish, sex control—through sex reversal technology, improved feeds, health management and bioremediation (Danish et al. 2017).

7.6.1   Low-tech Applications

Probiotics in Fish Feed   Increased intensification of aquaculture has led to higher disease outbreaks, with bacterial diseases being a major cause of economic losses and constraining its expansion (Pieters et al. 2008). The widespread use of chemotherapeutics/antibiotics to combat diseases has resulted in drug resistance in aquaculture (Nayak et al. 2007) and biosafety concerns, which have led to research for finding other methods of preventing losses (Rollo et al. 2006). Probiotics, live microbial supplements in feed with beneficial effects on the host[1], has been defined more broadly as 'a live microbial adjunct which has a beneficial effect on the host by modifying the host associated or ambient microbial community, by ensuring improved use of the feed or enhancing its nutritional value, by enhancing the host response towards disease, or by improving the quality of its ambient environment' (Verschuere et al. 2000). Probiotics are increasingly applied across the world as an alternative health management (Panigrahi et al. 2010). The use of probiotics as feed supplements started in the 1970s in pigs, cattle and poultry (Tukmechi et al. 2007) but is relatively new in aquaculture. Strains of lactic acid bacteria, such as *Lactobacillus*, have been administered to fish feed (Bucio-Galindo et al. 2009). Apart from its use as an antibiotic, probiotics have been applied in promoting growth in aquaculture and to improve nutritional composition (increasing crude protein, crude lipid) of Nile tilapia (*Oreochromis niloticus*), as well as increasing weight in the fish from 0.1 g to 6 g in 9 weeks of culture (Lara et al. 2003).

Although aquaculture is expanding in SSA, the development and application of probiotics are meagre when compared to other countries (Nwogu et al. 2011), perhaps constrained by a limitation on commercial probiotic products for aquaculture in the SSA market, the slow pace of screening for novel probiotic strains from local aquaculture rearing systems to suit the specific requirement and safety perceptions arising from their use. The majority of those using probiotics in SSA use the high-performing tilapia and catfish brood stock feed—e.g. EFICO Genio 838F (BioMar France)—which includes the probiotic Bactocell® and immune-modulating ingredients to improve survival and boost the immune system.

Use of probiotics in aquaculture systems across SSA is mainly confined to a few countries in the West African belt (Sierra Leone, Liberia, Senegal and Nigeria), Kenya in East Africa and Zambia in southern Africa. In Nigeria, for example, the ability of *Lactobacillus fermentum* (LbFF4), isolated from Nigerian *fufu*, a fermented, starchy staple food, and *L. plantarum* (LbOGI), isolated from a fermented beverage (*ogi*), have been evaluated for induction of immunity in African sharp-tooth catfish (*Clarias gariepinus*, Burchell) against selected fish bacterial pathogens (Ogunshe and Olabode 2009). Although the level of adoption by farmers

---

[1] A Joint FAO/WHO Expert Consultation on the Health and Nutritional Properties of Powder Milk with Live Lactic Acid Bacteria; Subsequently an expert working group in 2002 developed Guidelines for the Evaluation of Probiotics in Food, to identify and define what data need to be available to accurately substantiate health claims. See https://www.fao.org/food/food-safety-quality/a-z-index/probiotics/en/.

is still low, the level of research into this technology is increasing in Senegal, Nigeria and Kenya.

Sex Reversal Sex reversal is a process for changing a male to become female, or vice versa. The rationale is that male or female fish may have a faster growth rate. Sex-reversed fish populations may grow up to two to three times faster, thereby reducing the cost of production. Further, mixed-sex culture for species such as the Nile tilapia (*Oreochromis niloticus*) results in lack of uniformity, overcrowding and stunting (Omitoyin et al. 2013).

The Nile tilapia, for example, is a maternal mouthbrooder, where the female incubates the fertilized eggs in her buccal cavity until hatching. Embryos hatch 5–6 days post fertilization and start free swimming about a week after hatching. Gonad differentiation in the species occurs at 9–15 days post fertilization. It has been shown that temperature and steroid hormones can override genetic sex determination in this species at 2–3 and again 10–20 days post fertilization (Baroiller et al. 2009).

The most common sex reversal practice in aquaculture is hormonal induction of sex, using 17α-methyltestosterone (MT) to induce maleness, while reversal to female is achieved with estradiol-17β hormones via oral administration incorporated into a starter fish feed (Popma and Green 1990). However, other practitioners use pressure shock for reversing sexes, and androgenesis. Production of androgenetic diploid individuals (where all chromosomes of embryo are of paternal origin) is based on inactivation of the genetic materials of the egg, fertilization by normal sperm and blocking of the first mitotic division of the male origin haploid zygote. Androgenesis has been used in fish to produce males at the University of Stellenbosch (RSA), where an innovative YY technology (Brink et al. 2002) has been developed for producing all male progeny in the Nile tilapia.

Use of sex reversal is widespread in SSA, practiced in western Africa (Gabon, Sierra Leone, Nigeria, Liberia and Senegal), eastern Africa (Rwanda, Uganda, Tanzania, Kenya, Madagascar and Ethiopia) and southern Africa (Malawi and Zambia). In Tanzania, sex reversal is mostly implemented through commercial feeds pre-treated with hormones, a system that local researchers have been testing (e.g. Mbiru et al. 2015) for adoption by farmers. In these tests, feeds are treated with 17 α -MT hormone obtained from the Norwegian School of Veterinary Science (NVH). However, these methods are limited for local fish farmers in Tanzania, thereby limiting the expansion of aquaculture industry (Carlberg et al. 2000).

Sex reversal for monosex male tilapia breeding is practised in floating cages in Madagascar (FAO 2005). So widespread is the practice that the government issued an Interdepartmental Order (No.22914/2004 of 29/11/04) instituting a technical committee to promote monosex male tilapia in Madagascar. This committee came up with procedures for use of hormones for sex reversal. At the research level, FOFIFA has been optimizing technical procedures for producing monosex male tilapias.

In Kenya, more than half of the 87 government hatcheries are practising sex reversal using androgen-treated feeds and super YY technology, although success rate is reported to be suboptimal due to limited knowledge of the technology among the practitioners (Mirera et al. 2021). Reports from the field across Africa have shown sex reversal through androgenesis to increase yields in many culture systems thereby boosting food production from aquaculture (Jamu and Brummett 2002).

Polyploidy Induction Polyploidy is a condition in which the cells of an organism have more than two paired sets of chromosomes. Spontaneous polyploids are occasionally found in nature from wild and farmed species and appeared repeatedly during the evolution of fish. Because of this, the EU regulations (Directive 90/220/ CEE of April 23 of 1990) do not consider even artificially induced polyploids to be GMOs. Chromosome manipulation to develop polyploids and sterility has been demonstrated to improve growth and survival rates and increase quality of final products in aquaculture (Donaldson and Devlin 1996). Conceptually, sterile individuals divert their reproductive energy to growth. Most cultured species with simple reproductive strategy can readily be made triploids. Pressure and temperature shocks or crossing tetraploids with diploids are some of the methods used to achieve the desired ploidy level, depending on species.

Triploids have been produced by preventing the second meiotic division after the sperm enters the egg, thereby resulting in two sets of chromosomes from the female and one set from the male (Cherfas et al. 1993). This procedure is usually accomplished through chemical method, heat shock, cold shock, electric shock or hydrostatic pressure. Apart from increased growth rates, sterile triploids maintain the genetic diversity of native populations because of higher ploidy level, and where there is escape into the wild, sterility would prevent their establishment. Application of polyploidy is restricted to a few countries in SSA, with only Uganda, Kenya, Nigeria, Senegal and Burkina Faso applying the technology in the Nile tilapia.

Phenotypic Selection Phenotypic selection for improving desired traits and to aid crossbreeding (hybridization) is practiced in virtually all farmed organisms, including fish, as a means of improving economically important traits. Selection based on phenotypes in aquaculture is now a common practice across SSA. This is done to improve desired traits such as fast growth, weight and colour and is key to other practices such as hybridization/crossbreeding. Selective breeding has been used for the last two decades in Cote d' Ivoire, Ghana and Malawi to improve local strains of *O. niloticus* and *O. shiranus*, in collaboration with WorldFish (Gupta et al. 2004).

In Tanzania, fish hybrids have been produced by pairing *O. niloticus* females and *O. urolepis* hornorum males (Mbiru et al. 2015). In Ivory Coast, *Clarias gariepinus* have been hybridized with *Heterobranchus longifilis* (Legendre et al. 1992). Hybridization in catfish culture has been implemented in Nigeria (Nwadukwe 1995; Ayinla and Nwadukwe 2003) by crossing male *Heterobranchus bidorsalis* with *C. gariepinus*, as well as hybridization of *Claria anguillaris* and *H. longifilis* (Aluko 1995) so as to improve the commercial value by fast growth. Other examples include tilapia hybridization in Kenya (Agnese et al. 1998), where female *Tilapia tholoni* were crossed with a male *Oreochromis mossambicus*, yielding 100% females, but crossing female *O. spirilus* with male *O. leucostictus* yielded 98% males. In Kenya, the Kenya Marine and Fisheries Research Institute (KEMFRI) is engaged in Nile tilapia selective breeding; however, the long-term genetic improvement programme lacks skilled personnel (Munguti et al. 2021). Kenya has a total of 87 hatcheries spread across 31 regions (counties) for Nile tilapia and African catfish.

7.6.2 Medium-tech Applications

Medium-tech applications in aquaculture include MAS as well as PCR as a tool for various applications such as characterization and disease diagnostics.

Marker-Assisted Selection In the simplest terms, MAS is the selection for desired and economically important traits in breeding programmes by use of markers linked to genes which are associated with the traits. This association can either come through physical linkage (linkage disequilibrium) or through selection. MAS is a more precise, genotype-based, rather than phenotype-based selection.

The relative advantage of MAS over conventional breeding may depend on the traits targeted for improvement: MAS offers more advantage for traits that are difficult to record in the candidates, such as disease resistance, fillet quality and feed efficiency. It is expected that MAS would more rapidly be adopted compared to genetic engineering, since public acceptance of this technology is as high as for conventional breeding and regulation—from research to commercialization—is less stringent.

The application of MAS can be enhanced by molecular marker maps that have been constructed for a number of aquaculture species such as tilapia, catfish, prawns and Atlantic salmon. However, despite the existence of these linkage maps in economically important species usually cultured in SSA, such as tilapia, it appears to have only been used in Nigeria where MAS is used in African sharp-tooth catfish breeding (Ikpeme et al. 2015).

Polymerase Chain Reaction The development of the PCR has permitted rapid in vitro DNA amplifications. Consequently, molecular marker genotyping and DNA sequencing have proved one of the most popular methods for analyses of genetic variation, evolutionary relationships (Avise 2000) and characterization for conservation. The application of PCR at research level as a tool for studying genetic variation appears to be widespread across SSA, usually happening at most institutions that have a molecular biology laboratory.

Cases where PCR is applied in fisheries and aquaculture as a tool include assessment of stock structure in Spanish mackerel (*Scomberomorus commerson*) in the northern Tanzania coastal waters (Johnson et al. 2017) and its utilization in MAS selection and breeding for African sharp-tooth catfish (*C. gariepinus*) aquaculture in Nigeria (Ikpeme et al. 2015). Further, there have been several other studies for sexing shrimps in Mozambique, and with a conservation focus in Madagascar, Tanzania, Sierra Leone, Nigeria, Uganda, Kenya, Malawi and Zambia.

7.6.3   High-tech Applications

Technologies classified as high-tech for this analysis include the broad area of
genetic engineering, including gene editing.

Genetic Engineering Advances in biotechnology and better understanding of
genomes have provided the tools necessary for manipulation of genes and
chromosomes in living organisms. Genetic engineering refers to a wide range of
manipulations of the genome of an organism. Some of these manipulations result in
GMOs, while others do not. For example, manipulations that terminate processes or
pathways, such as RNAi, should not be regarded as GMOs. FAO defines GMOs as
those that have had foreign DNA artificially inserted into their own genomes. As
such, GMOs are necessarily transgenics. Many genetically modified fish species
have been produced through various methods for incorporating the foreign gene,
including microinjection, electroporation, infection with pantropic defective retrovi-
ral vectors, particle gun bombardment and sperm- and testis-mediated gene transfer
methods (Rasmussen and Morrissey 2007).

Genetic engineering can be applied to achieve sex reversal in fish, where it is used
to link the growth hormone gene to the alpha-fetoprotein (AFP) gene promoter as a
way of bypassing the central nervous system control on growth hormone expression.
This is expected to improve growth rates. This has been done for several fish species,
including goldfish, mud loach (Zhu et al. 1986) and tilapia (*Oreochromis niloticus*)
(Maclean and Laight 2000) using a growth promoter from salmon. However, the
current controversy surrounding the adoption of foods derived from genetic engineering is protracted, reminiscent of events that occurred during the Green Revolution more than half a century ago. A typical example is the hostility with which
genetically improved fish has been received in parts of SSA. In Nigeria, for example,
hostility is reported towards genetically improved fish and products because some
people erroneously believing that such fish will cause cancer (Erondu et al. 2011).
There are no documented cases where genetic engineering has been applied to the
improvement of either wild or farmed fish in SSA.

Gene Editing Gene editing technologies are relatively new, with a recent attempt
being directed at successfully altering the coat colour of sheep using clustered
regularly interspaced short palindromic repeats (CRISPR-Cas9) by a team of
scientists in China (Zhang et al. 2017). Compared with traditional gene mutation
approaches, in which researchers take decades to breed a new strain, gene editing can
yield much faster results. This technology has also been successfully applied in basic
science to study the evolution of body parts in fishes.

Gene editing intervenes in the genetic material by modifying the genome of the
organism, but in contrast to gene transfer, it usually only manipulates the species-
specific genome. Special techniques are used to delete or replace sections of a gene.
Among the latest promising gene editing technologies for aquaculture, one, known
as the CRISPR/Cas9, is noteworthy here. The procedure is a kind of 'gene scissors'
with which the hereditary molecule DNA in the fertilized fish egg can be cut in a

specific place (usually within the desired target gene). The cell recognizes the inflicted injury and repairs the damage independently. Since after the repair the DNA sequence often differs from the original version, this method can be used to disrupt or prevent the expression of targeted functional proteins. In contrast to gene transfer, gene editing does not involve the transfer of foreign DNA from other organisms and its integration into the target genome and hence does not produce a transgenic organism. This method *could* therefore stand a better chance of public acceptance. CRISPR/Cas9 technology has already been applied successfully in some fish species in aquaculture, e.g. salmon, rainbow trout, catfish, tilapia and carp. For example, Nakamura et al. (2016) used gene editing techniques to conduct functional analysis using CRISPR/Cas9 and fate mapping in zebrafish to unravel the development of the wrists and digits of tetraploids. By knocking out the hox13 gene in fish, they found an association between reduction and loss of fin rays and increased number of endochondral distal radials. Gene editing has also been applied successfully in making sterile salmon, using CRISPR technology in a project at Institute of Marine Research of Norway, seeking to protect the genetic integrity of wild salmon and trout. The project is now applying the technology to improve disease resistance and nutritional profile by producing more omega-3. There are no known cases of application of gene editing in improvement of either wild or farmed fish in SSA.

Gene editing can also be used for precision breeding. For example, it can be used to specifically implement the necessary genetic sequences in fish populations that lack certain desirable traits. This is done without affecting the other important functional traits, something very difficult to achieve through conventional breeding methods.

These biotechnologies and the main countries in which they are applied are summarized in Table 7.4.

7.7   Classification of Countries on the Basis of Biotechnology Applications in Aquaculture

Two areas in aquaculture, not related to use of specific biotechnologies, but which indicate country engagements in aquaculture as an important economic sector, need mention here. These are seaweed culture and marine shrimp farming.

Seaweed Culture Seaweed is the common name for a large number of species of marine macrophytic plants and algae that grow in oceans, rivers, lakes and other water bodies, generally in the shallow waters in the tidal zone. Some seaweeds are microscopic, such as the phytoplankton which live as suspended bodies in the water column, but others are large, e.g. the giant kelp that grows in 'forest-like form' from their roots at the bottom. Seaweeds exhibit highest photosynthesis efficiency due to moist conditions and are considered to contribute about 50% of all photosynthesis in the world. Seaweed farming is the practice of cultivating and harvesting seaweed. It consists, in the simplest form, of the management of naturally found seaweed and, in the most advanced form, of fully controlling the life cycle of the plants and algae.

Table 7.4 Examples of countries applying specific technologies (in research and at farm level) in aquaculture in SSA

Biotechnology		Countries
Low-tech	Probiotics in fish feed	*Limited use*—Kenya, Liberia, Nigeria, Senegal, Sierra Leone and Zambia
	Sex reversal using growth hormones in feed	*Widespread use*, including: Ethiopia, Gabon, Kenya, Liberia, Madagascar, Malawi, Nigeria, Rwanda, Senegal, Sierra Leone, Tanzania, Uganda and Zambia
	Polyploidy induction	*Limited use*—Burkina Faso, Kenya, Nigeria, Senegal and Uganda
	Phenotypic selection	*Widespread use*, applied in virtually all countries where aquaculture is practiced
Medium-tech	Marker-assisted selection (MAS)	*Nigeria*
	PCR—as a tool	*Widespread use*, including: Kenya, Madagascar, Malawi Mozambique, Nigeria, Sierra Leone, Tanzania, Uganda and Zambia
High-tech	Genetic engineering	*None*; unconfirmed reports of use in Senegal, Tanzania and Zambia
	Gene editing	*None*; unconfirmed reports of use in Senegal

Seaweeds are rich in vitamins and minerals and are consumed as food in various parts of the world and used for the production of phytochemicals, e.g. agar, carrageenan and alginate, which are widely used as gelling. Seaweed cultivation has potential as a diversification activity in mariculture. Seaweeds also help in soaking up what is in the water and can thus potentially be cultivated to reduce heavy metals and other water pollutants. The practice of seaweed culture in SSA is currently limited to Mozambique, South Africa, Namibia and Tanzania.

Shrimp Culture The global demand for shrimps is estimated at about 4.75 million tonnes annually. There has been a shortfall in meeting this demand because more than 50% of the shrimps sold worldwide are obtained from the wild. China and Thailand are the world's leading producers of shrimps, accounting for nearly 75% of the global shrimp production. In 2018, Thailand was the biggest shrimp producer with shrimp production of 2.59 million metric tons. The other 25% is produced mainly in Latin America, where Brazil, Ecuador and Mexico are the largest producers. The largest exporting nation is India. Technological advances have led to growing shrimp at increasingly higher densities, and brood stock is shipped worldwide. The commercial culture of marine shrimp in tropical areas has also grown rapidly during the last two decades. However, only a few SSA countries have taken up marine shrimp farming. Marine shrimp culture in SSA is concentrated in Madagascar, although there are a few farms in Seychelles, Kenya and Mozambique. There have also been attempts at raising shrimp in Gabon and Nigeria. Molluscs culture is limited to Namibia and South Africa, and the latter also has regionally significant production of mussels and oysters and has initiated production of abalone (ear shell).

Table 7.5 Classification of countries on the basis of biotechnology applications in aquaculture

Applications category	Countries
Very low use	Angola, Benin, Botswana, Burundi, Cameroon, CAR, Chad, Congo, Côte d'Ivoire, Djibouti, DRC, Equatorial Guinea, Eritrea, Eswatini, Ethiopia, Gambia, Ghana, Guinea, Guinea-Bissau, Lesotho, Liberia, Mali, Mauritius, Namibia, Niger, Rwanda, Senegal, Somalia, South Sudan, Togo, Zambia
Low use	Burkina Faso, Gabon, Malawi, Mozambique, Sudan, Uganda, Zimbabwe
Medium use	Kenya, Madagascar, Nigeria, Sierra Leone, South Africa, Tanzania
High use	None
Very high use	None

Although it would generally be expected that application of low-level technologies is widespread, both in research and commercially by farmers, while medium-tech and high-tech technologies would be restricted to research, the pattern of application for some of the technologies in SSA does not necessarily reflect this: Whereas the use of probiotics in feed is relatively simple and cost-effective compared to hormone-based sex reversal, we found the use of probiotics to be limited to only a few countries, while farm application of sex reversal was widespread across the SSA region. This contrast points to a pattern where technology adoption at farm level is driven by the need to reduce production costs. Indeed, comparing the three low-tech technologies analysed, sex reversal was the most widespread at farm level.

Table 7.5 shows that, for the 44 countries with information, 38 were in the 'very low' (31) or 'low' use (7) categories and six countries (Kenya, Madagascar, Nigeria, Sierra Leone, South Africa and Tanzania) in the 'medium' category. None were in the 'high' or 'very high' categories.

Countries not listed due to lack of sufficient data: Cape Verde, Comoros, Mauritania, Sao Tome and Principe and Seychelles

7.8   Biotechnology Impact Case Studies in Aquaculture

7.8.1   Monosex Tilapia

The Nile tilapia, *Oreochromis niloticus*, is an African freshwater fish from the cichlid family which naturally occurs in the Nile Basin. Although historical records suggest that this species was cultivated in Egypt more than 3000 years ago, modern fish farming really took off from the second half of the twentieth century. Today, Nile tilapia are farmed in warm water areas throughout the world with the vast majority being reared outside Africa, especially in Asia.

Nile tilapia have many features that make them good candidates for aquaculture: they are tolerant of a wide range of conditions; can convert largely plant-based diets into high-quality flesh; are suitable for polyculture with other fish, including African catfish; and are easy to fillet as there are no bones in the muscle.

The major disadvantage of Nile tilapia is that when kept in mixed sex groups they are very prolific breeders. Female fish direct their energy towards reproduction rather than growth, and the result is large numbers of small fish in the rearing ponds. These fish are too small to attract good prices in the market.

The solution to this problem is to rear just males, which can reach 500 g or more in less than a year. Several methods have been developed to enable this including hand sexing, but the most commonly used method involves sex reversal using a synthetic, commercially available steroid hormone, 17α-methyltestosterone. Newly hatched fry are fed on high-quality feed that contains the hormone for between 25 and 28 days. Depending on how well this approach is applied, 95–100% of the fry are male. Compared to mixed sex ponds, Nile tilapia in monosex ponds grow faster, achieve up to 50% higher live weights with greater uniformity and have better feed conversion ratios.

In 2018, FAO reported that there were 125 active fish hatcheries in Kenya, more than 80% of which were privately owned with 5 private hatcheries accounting for around 90% of fingerlings produced. The majority of hatcheries (more than 90%) had not implemented monosex production citing financial constraints to purchase the sex reversal hormone and lack of the required infrastructure: brood fish and fry need to be reared in separate ponds or cages within ponds to implement the sex reversal process. The larger-scale hatcheries produced both mixed sex and monosex tilapia. Mixed sex fingerlings are cheaper than monosex fingerlings: in 2017, while the former were sold for the equivalent of USD 0.036 each, the latter were USD 0.06.

In Ghana, use of monosex production was reported to be much more common in floating cage systems than ponds. Profit per hectare was also reported to be 2.7 times for monosex versus mixed sex production. Meanwhile, use of monosex production was reported to be well established in Egypt and Nigeria.

Sources: Fregene et al. (2020), Mlalila et al. (2015), Omasaki et al. (2016), Nyonje et al. (2018), Kalima (2020), Snake et al. (2020).

References

Agnese JF, Adepo-Gourene B, Pouyoud L (1998) Natural hybridization in tilapias. In: Agnese JF (ed) Genetics and aquaculture in Africa. ORSTOM Editions, pp 97–103

Aluko PO (1995) Growth characteristics of first, second and backcross generations of the hybrids between Heterobranchus longifilis and Clarias anguillaris. 1995 Annual Report. National Institute for Freshwater Fisheries Research, New Bussa, pp 74–78

Avise JC (2000) Phylogeography: the history and formation of species. Harvard University Press, Cambridge, MA

Ayinla OA, Nwadukwe FO (2003) Review of the development of hybrid (Heteroclarias) of Clarias gariepinus and Heterobranchus bidorsalis. Nigeria J Fish 1:85–98

Baroiller JF, D'Cotta HH, Bezault E et al (2009) Tilapia sex determination: where temperature and genetics meet. Comp Biochem Physiol A Mol Integr Physiol 153(1):30–38

Brink D, Mair GC, Hoffman L et al (2002) Genetic improvement and utilisation of indigenous tilapia in southern Africa: final technical report, December 1st 1998 to June 31st, 2002

Brummett RE, Lazard J, Moehl J (2008) African aquaculture: realizing the potential. Food Policy 33(5):371–385. https://doi.org/10.1016/j.foodpol.2008.01.005

Bucio-Galindo A, Hartemink R, Schrama JW et al (2009) Kinetics of Lactobacillus plantarum 44a in the faeces of tilapia (Oreochromis niloticus) after its intake in feed. J Appl Microbiol 107: 1967–1975

Cai J, Quagrainie K, Hishamunda N (2017) Social and economic performance of tilapia farming in Africa. FAO Fisheries and Aquaculture Circular (C1130)

Carlberg JM, Olst JVC, Massingi MJ (2000) Hybrid striped bass: an important fish in US aquaculture. Aquac Mag 26:26–38

Cherfas NB, Gomelskey BI, Peretz Y et al (1993) Induced gynogenesis and polyploidy in the Israel common carp line Dor-70. Isr J Aquacult 45:59–72

Danish M, Trivedi RN, Kanyall P et al (2017) Importance of biotechnology in fish farming system: an overview. Progress Res J 12(1):7–14

Delgado CL, Wada N, Rosegrant MW et al (2003) Fish to 2020: supply and demand in changing global market. International Food Policy Research Institute and WorldFish Center, Washington, DC

Donaldson EM, Devlin RH (1996) Uses of biotechnology to enhance production. In: Pennel W, Barton BA, Pennell WA (eds) Principles of salmonid culture, vol 29. Elsevier Science, Amsterdam, pp 969–1020

Erondu ES, Akinrotimi OA, Gabriel UU (2011) Genetic manipulation for enhanced aquaculture production in Nigeria. Biosci Res 23(2):131–140

FAO (2005) National aquaculture sector overview madagascar. In: Fisheries and aquaculture division. Available via FAO: https://www.fao.org/fishery/countrysector/naso_madagascar/en. Accessed 30 Aug 2021

FAO (2015) FAO aquaculture Newsletter 53. Available via FAO: https://wwwfaoorg/documents/card/en/c/b6c21151-48f6-49f9-8749-75e155325fe6/. Accessed 30 Aug 2021

FAO (2020) The State of world fisheries and aquaculture (SOFIA) 2020. Sustainability in action. FAO, Rome, pp 244. http://www.fao.org/3/ca9229en/online/ca9229en.html. Accessed 30 Aug 2021

Fregene BT, Karisa HC, Olaniyi AA et al (2020) Extension manual on mono-sex Tilapia production and management (First Edition): technologies for African Agricultural Transformation (TAAT) aquaculture compact, WorldFish, Penang. Available via WorldFish. https://hdl.handle.net/20. 500.12348/4742

Gupta MV, Bartley DM, Acosta BO (eds) (2004) Use of genetically improved and alien species for aquaculture and conservation of aquatic biodiversity in Africa. World Fish Center Conf Proc 68: 113

Halwart M (2020) Fish farming high on the global food system agenda in 2020. FAO Aquaculture Newsletter 61:11–111

Hecht T, Moehl JF, Halwart M et al (2006) Regional review on aquaculture development. 4. Sub-Saharan Africa. Fisheries Circular. No. 1017/5. FAO, Rome, p 97

Ikpeme EV, Udensi OU, Ekaluo UB et al (2015) Unveiling the genetic diversity in Clarias gariepinus (Burchell, 1822) using random amplified polymorphic DNA (RAPD) fingerprinting technique. Asian J Anim Sci 9(5):187–197

Jamu D, Brummett R (2002) Opportunities and challenges for African aquaculture. Expert consultation on biosafety and environmental impact of genetic enhancement and introduction of improved tilapia strains/alien species in Africa, Nairobi

Johnson MG, Mgaya YD, Shaghude YW (2017) Mitochondrial DNA control region revealed a single genetic stock structure of scomberomorus commerson lacepede (1800) in the Northern Tanzania Coastal Waters. IOTC-2017-WPNT07-16. Available via IOTC. https://www.iotc.org/documents/mitochondrial-dna-control-region-revealed-single-genetic-stock-structure-scomberomorus. Accessed 25 Oct 2021

Kalima S, Jere WL, Kefi AS (2020) Growth performance of monosex and mixed sex of Oreochromis tanganicae (Günther, 1894) raised in semi concrete ponds. J Aquac Res Dev 11(4):3–16. https://doi.org/10.35248/2155-9546.19.11.586

Lara F, Olvera N, Guzm'an M et al (2003) Use of the Bacteria streptococcus faecium and lactobacillus acidophilus, and the yeast saccharomyces cerevisiae as growth Promoters in Nile Tilapia (Oreochromis Niloticus). Aquac 216:1–4

Legendre M, Teugels GG, Cauty C et al (1992) A comparative study on morphology, growth rate and reproduction of Clarias gariepinus, Heterobranchus longifilis and their reciprocal hybrids. J Fish Biol 40:59–79

Machena C, Moehl J (2001) Sub-Saharan African aquaculture: regional summary. In: Subasinghe RP, Bueno P, Phillips MJ et al (eds) Aquaculture in the third millenium. Technical Proceedings of the Conference on Aquaculture in the Third Millennium, Bangkok, Thailand, 20–25 February 2000. NACA, Bangkok and FAO, Rome, pp 341–355

Maclean N, Laight RJ (2000) Transgenic fish: an evaluation of benefits and risks. Fish Fish 1:146–172

Mbiru M, Limbu SM, Chenyambuga SW et al (2015) Comparative performance of mixed-sex and hormonal sex-reversed Nile tilapia Oreochromis niloticus and hybrids (Oreochromis niloticus 3 Oreochromis urolepis hornorum) cultured in concrete tanks. Aquac Int 24(2):557–566

Mirera D, Nyonje B, Opiyo M et al (2021) Aquaculture research and training in Kenya. In: Munguti J, Obiero K, Orina P et al (eds) State of aquaculture report 2021: towards nutrition sensitive fish food production systems. Techplus Media House, Nairobi, p 190

Mlalila N, Mahika C, Kalombo L et al (2015) Human food safety and environmental hazards associated with the use of methyltestosterone and other steroids in production of all-male tilapia. Environ Sci Pollut Res 22(7):4922–4931. https://doi.org/10.1007/s11356-015-4133-3

Munguti JM, Kim JD, Ogello EO (2014) An overview of Kenyan aquaculture: Current status, challenges, and opportunities for future development. Fish Aquatic Sci 17(1):1–11. https://doi.org/10.5657/FAS.2014.0001

Munguti J, Obiero K, Orina P et al (eds) (2021) State of Aquaculture Report 2021: towards nutrition sensitive fish food production systems. Techplus Media House, Nairobi, p 190

Nakamura T, Gehrke AR, Lemberg J et al (2016) Digits and fin rays share common developmental histories. Nature 537(7619):225–228

Nayak SK, Swain B, Mukherjee SC (2007) Effect of dietary probiotic and vitamin C on the immune response of India major carp Labeo rohita (Ham). Fish Shellfish Immunol 23:892–896

Nwadukwe FO (1995) Hatchery propagation of five hybrid groups by artificial hybridization of Clarias spp. and Heterobranchus spp. using dry powdered carp pituitary hormone. J Aquacul Trop 10:1–10

Nwogu NA, Olaji ED, Eghomwanre AF (2011) Application of probiotics in Nigeria aquaculture: current status, challenges and prospects. Int Res J Microbiol 2(7):215–219

Nyandat B, Owiti G (2013) Aquaculture needs assessment mission report. Report/Rapport: SF-FAO/2013/24. September/Septembre 2013. FAO-SmartFish Programme of the Indian Ocean Commission, Ebene

Nyonje BM, Opiyo MA, Orina PS et al (2018) Current status of freshwater fish hatcheries, broodstock management and fingerling production in the Kenya aquaculture sector. Livest Res Rural Dev 30(1):1–8 Available via LRRD: http://wwwlrrdorg/lrrd30/1/mary30006html. Accessed 26 Oct 2021

Ogunshe AA, Olabode PO (2009) Antimicrobial potentials of indigenous lactobacillus strains on gram – negative indicator bacterial species from Clarias gariepinus (Burchell) microbial inhibition of fishborne pathogens. Afr J Microbiol Res 3(12):870–876

Omasaki SK, Charo-Karisa H, Kahi AK et al (2016) Genotype by environment interaction for harvest weight, growth rate and shape between monosex and mixed sex Nile tilapia (Oreochromis niloticus). Aquac 458:75–81. https://doi.org/10.1016/j.aquaculture.2016.02.033

Omitoyin BO, Ajani EK, Sadiq HO (2013) Preliminary investigation of Tribulus terrestris (Linn. 1753) extracts as natural sex reversal agent in Oreochromis niloticus (Linn. 1758) larvae. Int J Aquac 3:133–137

Panigrahi A, Kiron V, Satoh S (2010) Probiotic bacteria Lactobacillus rhamnosus influences the blood profile in rainbow trout Oncorhynchus mykiss (walbaum). Fish Physiol Biochem 36:969–977

Pieters N, Brunt J, Austin B et al (2008) Efficacy of in-feed probiotics against Aeromonas bestiarum and Ichthyophthirius multifiliis skin infections in rainbow trout (Oncorhynchus mykiss, Walbaum). J Appl Microbiol 105:723–732

Polidoro B, Ralph GM, Strongin K et al (2016) Red list of marine bony fishes of the Eastern Central Atlantic, Gland, p 80. https://doi.org/10.2305/IUCN.CH.2016.04.en

Popma TJ, Green BW (1990) Sex reversal of tilapia in earthen ponds. Research and development series. International Center for Aquaculture, Alabama Agriculture Experiment Station, Auburn University, Auburn, AL, p 15

Rasmussen RS, Morrissey MT (2007) Biotechnology in aquaculture: transgenics and polyploidy. Compr Rev Food Sci Food Saf 6:2–16. https://doi.org/10.1111/j.1541-4337.2007.00013.x

Rollo A, Sulpizio R, Nardi M et al (2006) Live microbial feed supplement in aquaculture for improvement of stress tolerance. Fish Physiol Biochem 32:167–177

Rukanda JJ (2018) Evaluation of aquaculture development in Tanzania. Nations University Fisheries Training Programme, Iceland [final project]. http://www.unuftp.is/static/fellows/document/janeth16aprf.pd

Snake M, Maluwa A, Zidana H et al (2020) Production of a predominantly male tilapia progeny using two Malawian tilapias, Oreochromis shiranus and Oreochromis karongae. Aquac 16:100274. https://doi.org/10.1016/j.dib.2020.105716

Tukmechi A, Morshedi A, Delirezh N (2007) Changes in intestinal microflora and humoral immune response following probiotic administration in rainbow trout (Oncorhynchus mykiss). J Anim Vet Adv 6(10):1183–1189

Uwagbale E (2021) How Nigeria became the world's largest producer of African catfish. Available at: https://scitechafrica.net/how-nigeria-became-the-worlds-producer-of-afri can-catfish/. Accessed 7 Sept 2021

Verschuere L, Rombault G, Sorgeloos P et al (2000) Probiotics bacteria as biological control agents in aquaculture. Microbiol Mol Biol Rev 64:655–671

Zhang X, Li W, Liu C et al (2017) Alteration of sheep coat color pattern by disruption of ASIP gene via CRISPR Cas9. Sci Rep 7:8149. https://doi.org/10.1038/s41598-017-08636-0

Zhu Z, Xu K, Li G et al (1986) Biological effects of human growth hormone gene microinjected into the fertilized genes of loach Misgurnus anguillicaudatus (Cantor). Chin Sci Bull 31(41):387–389

Overall Status, Gaps and Opportunities for Agricultural Biotechnology in Sub-Saharan Africa

8

J. E. O. Rege and Keith Sones

Abstract

Combining the assessments for capacity, the enabling environment and application for all four sectors, only two countries, Kenya and South Africa, are considered to have an overall agricultural biotechnology status of 'strong'. Eleven countries fall in the 'medium' category and 30 are categorised as 'weak'. Sixteen of the 'weak' countries are highly dependent on agriculture. Without the benefits that can be delivered by modern biotechnologies, it is unlikely that these countries will achieve food security and provide sustainable livelihoods for their rapidly growing populations. In contrast, eight countries whose economies are highly dependent on agriculture have 'medium' or 'strong' overall biotechnology status. These countries appear to have high potential to reap the benefits of modern and advanced biotechnologies. Ten critical gaps were identified, each with corresponding opportunity areas for action with regard to agricultural biotechnologies in sub-Saharan Africa. They are human capacity, especially lack of critical mass in advanced technologies; research infrastructure, facilities and equipment, including establishment of shared facilities; financial resources; policies and biosafety frameworks; taking research products in to use; public-private partnerships; intellectual property protection; administrative structure and inter-sectoral coordination; awareness and public participation; and the low priority given to forestry and aquaculture across SSA.

J. E. O. Rege (✉)
Emerge Centre for Innovations-Africa, Nairobi, Kenya
e-mail: ed.rege@emerge-africa.org

K. Sones
Keith Sones Associates, Banbury, UK

© The Author(s), under exclusive license to Springer Nature Switzerland AG 2022   173
J. E. O. Rege, K. Sones (eds.), *Agricultural Biotechnology in Sub-Saharan Africa*,
https://doi.org/10.1007/978-3-031-04349-9_8

8.1   Introduction

This chapter uses the results of the analysis of agricultural biotechnology capacities, the enabling environment and applications across the four sectors—crops, livestock, forestry and aquaculture—to provide an overall picture of the status of sub-Saharan Africa (SSA) countries. The cross-sector analysis is first done separately on the basis of capacities, enabling environment and applications. These are then combined to provide an overall picture. Finally, the chapter examines current gaps in, and opportunities for, enhancing applications of agricultural biotechnology in SSA.

8.2   Capacity for Biotechnology Applications Across Sectors of Agriculture

Perhaps more than any factor, the state of a county's capacity affects the level to which agricultural biotechnologies can be applied and the pace of change which can be realized in the transformation of agriculture. In the last two decades, many SSA countries have made considerable efforts in the development of capacity for research, with a subset being more intentional on harnessing opportunities for the application of biotechnology. For instance, agricultural research spending and human resource capacity both grew in SSA as a whole during 2000–2014, but results were uneven, with a number of countries showing stagnating or declining investment growth.

Underinvestment in agricultural research continues. In most countries, the capacity development efforts made are insufficient to enable substantial benefits from biotechnological innovations. The region's agricultural research intensity ratio fell during 2000–2015 because growth in agricultural research spending was slower than growth in agricultural output over time. Furthermore, agricultural research spending became more dependent on volatile donor funding. Consequently, many SSA countries face serious human resource capacity and infrastructure challenges. As of 2014, a large number of agricultural researchers, especially those qualified to the PhD level, were approaching retirement age, representing a significant risk that the affected agencies could be left without the critical mass of senior, well-experienced researchers needed to lead research programmes. This trend has continued in the decade to 2019/2020 and, combined with high numbers of more recently recruited junior staff in need of experience and mentoring, has left many countries vulnerable. Without deliberate recruitment and adequate succession strategies and training, significant knowledge gaps will persist or emerge, raising concerns about the quality of future research outputs. Outdated research facilities and equipment, especially in the biosciences area, are also impeding the conduct of productive research, which compromises the number and quality of research outputs and ultimately translates into reduced impact.

Although there are no comprehensive data on R&D investments by countries, investments in capacities for biotechnology development and application remain generally inadequate. There is a chicken-and-egg situation here: the limited number of funded projects and project pipelines across SSA are directly related to inadequate capacities—with countries doing well in terms of capacities also doing well in well-funded projects. Indeed, countries with exceedingly low capacities tended to have only occasional time-bound projects linked to development grants or loans; such countries do not have the calibre of personnel to win competitive grants. In response to these capacity challenges, some governments are demonstrating what commitment can achieve, with Ethiopia as a case in point (see Box 8.1). These interventions by Ethiopia are beginning to pay off: as strong national agricultural research system capacity facilitates mobilization of additional resources, funders see the quality and relevance of research, and this, in turn, attracts more funding. Clearly, as is being demonstrated by Ethiopia, it is possible to break out of the current state of low capacity, no funding and no tangible and relevant research deliverables.

Box 8.1 Government Action Towards Addressing Agricultural Biotechnology Capacities: The Case of Ethiopia

The Ethiopian government has consistently invested in biotechnology infrastructure and facilities. This started in the mid-1990s with support to the Biodiversity Research Institute to establish biotechnology research facilities for tissue culture and molecular marker technologies.

The strong government support combined with a World Bank loan (as part of the East African Agricultural Productivity Programme (EAAPP)) increased Ethiopia's agricultural research spending by about 25% during 2011–2014 in inflation-adjusted terms. Ethiopia's pool of agricultural researchers was also expanded by about 900 full-time equivalents (FTEs) during the same period. This growth occurred evenly across the Ethiopian Institute of Agricultural Research (EIAR), the regional agricultural research institutes (RARIs) and higher education agencies.

In addition, EIAR and the RARIs received substantial funding through EAAPP and other donor-supported programmes to upgrade laboratory infrastructure and equipment; many laboratories benefited from this investment.

Recognizing the fact that the majority of researchers employed at EIAR and the RARIs only held BSc degrees and that turnover among MSc- and PhD-qualified researchers was high, the government, in 2014, doubled the salary levels of senior researchers employed at EIAR. In addition, donors have contributed USD1 million to enable the hiring of retired PhD-qualified researchers to mentor EIAR's young researchers.

Data from the IFPRI-led Agricultural Science and Technology Indicators (ASTI) initiative show that the total agricultural R&D spending in Ethiopia (constant 2011 purchasing power parity dollars) in 2014 was estimated at USD123.7 million (up from 56.6 m in 2000 and 104.1 m in 2010) and total number of researchers was 2768.5 FTEs (up from 743.8 in 2000 and 1680.5 in 2010). The spending in crop and livestock biotechnology in 2014 was estimated at USD15.2 million.

(continued)

Box 8.1 (continued)

As part of this concerted capacity development, the government, in 2015, established a National Agricultural Biotechnology Research Centre at Holeta, just outside Addis Ababa with an investment estimated at USD16 million. Currently, the centre conducts research in well-equipped laboratories focusing on plant tissue culture and molecular marker technology, as well as animal and microbial biotechnology. In addition, the government has fully supported the establishment of the National Veterinary Research Institute, a well-equipped and staffed facility, to carry out research in vaccine development. It produces vaccines against several animal diseases such as anthrax, black leg, contagious bovine and caprine pleuropneumonia, foot-and-mouth disease, Newcastle disease, rinderpest and sheep pox. This institute also undertakes research to develop animal diagnostics and preventative and treatment tools using modern biotechnologies such as DNA probes and immunoblotting. Biotechnology-related animal breeding techniques undertaken by the institute include artificial insemination and multiple ovulation and embryo transfer.

Source: Compiled by authors

An indication of the current state of capacities for biotechnology in SSA countries is summarized in Fig. 8.1 which presents total agricultural R&D staffing (obtained from ASTI data) as well as estimated biotechnology staffing. It should be noted that this covers mostly crops and livestock.

Ethiopia leads in total ARD FTEs as well as biotech content of the staffing. However, it must also be noted that the Ethiopian efforts are relatively recent and most of the staff are young and early career—predominantly BSc and MSc (see Box 8.1). Other high-ranked countries for biotechnology capacity are Nigeria, Kenya and South Africa. Mauritius, Botswana, Namibia and Cape Verde come out top when FTEs are expressed per million inhabitants, although Ethiopia still features prominently here too.

The FTEs discussed above reflect public sector investments in staffing in NARIs and universities. Another, and important, dimension is availability of relevant training opportunities in the country. Figure 8.2 is a summary of universities and colleges dedicated to, or with programmes on, agricultural biotechnology. Nigeria leads with 21 such institutions, followed by South Africa (20), Kenya (18), DRC (11) and Uganda (11). These figures do not necessarily speak to total student enrolments, nor to quality of training. They only indicate investments made towards training to support agricultural R&D. This is only one indicator—as Ethiopia (with 7 agricultural universities—the majority very young) has demonstrated in terms of agricultural R&D staffing.

Figure 8.3 summarizes the status of agricultural research personnel across sectors (in FTEs) in SSA countries (excluding South Africa whose data was not available) in absolute numbers and expressed per million inhabitants. In absolute numbers, Ethiopia, Nigeria, Kenya and Tanzania—all countries endowed with larger numbers

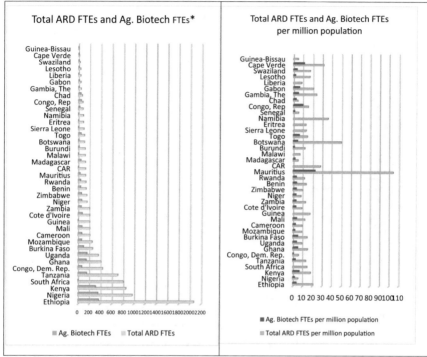

*Ag. Biotech FTEs calculated based on crop and livestock data only

Fig. 8.1 Agricultural R&D (ARD) and biotechnology staffing (FTEs) in SSA—absolute and per million inhabitants (2014). *Ag. Biotech FTEs calculated based on crop and livestock data only

of agricultural R&D personnel—are also among the countries found at the top in terms of applications. South Africa belongs in this cluster. In addition, the top ten include DRC, Ghana, Uganda, Burkina Faso and Mozambique. However, when total spending is expressed per million inhabitants, other countries emerge—namely, Mauritius, Botswana, Namibia, Cape Verde and Central African Republic.

Classification of countries on the basis of capacities (human, infrastructural/ facilities and institutional networks/collaboration) across all sectors is summarized in Table 8.1. Clearly, the current state of capacities for biotechnology across SSA should be of concern, with only four countries (Ethiopia, Kenya, Nigeria and South Africa) showing 'medium' or higher capacity, consistently across all four sectors. On the other hand, 21 SSA countries analysed were consistently in the 'very low' category, while 18 others were in the 'low' category, across all sectors.

The analysis found countries to vary significantly in their capacities across specific biotechnologies and overall, due to investments made to promote specific biotechnologies and to develop human capacity and infrastructure and involvement of the private sector. Across the sectors, the most widespread technologies being developed and applied across SSA were those aimed at increasing yields and reducing losses (disease/pest control). These are mostly low- to medium-level

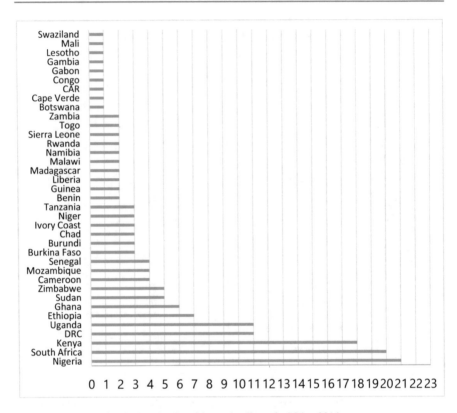

Fig. 8.2 Number of agricultural universities and colleges in SSA—2014

technologies, including genetic engineering for pest control in crops, diagnostics and vaccine development in livestock and sex reversal in fish. The most lacking capacity in human resources and infrastructure was that for developing and applying high-tech applications. Insufficient staffing, infrastructure and funding constrain not just the development of new technologies but also the adaptation and use of available and profitable biotechnologies already demonstrated to work elsewhere.

A trend that appeared across all sectors was the relatively higher capacity ranking of countries that host CGIAR and other international centres, and/or engage in international collaborations, have vibrant private sector involvement and have benefitted from capacity assistance by FAO, USAID, CIDA and other development agencies. Other factors that appear to enhance within-country capacities—and create an enabling environment (see Chap. 3) for research and applications—include public-private partnerships (PPPs), robust extension services and institutions that purposefully create or enhance awareness on agricultural biotechnologies.

Generally, crops provided a 'frontier' for capacity and development of biotechnologies which are then transferred to livestock and, to a lesser extent, forestry and aquaculture. This is consistent with the observation that most

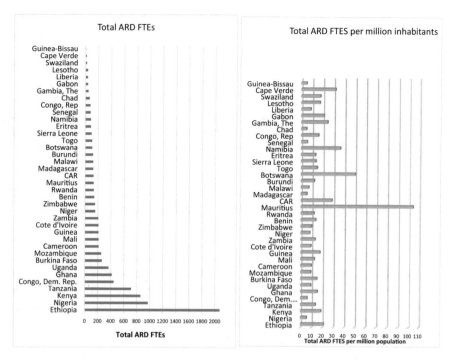

Fig. 8.3 Total R&D (FTEs) and ARD FTEs per million inhabitants

Table 8.1 Classification of countries across sectors on basis of capacities for biotechnology applications

Capacities across sectors	Countries[a]
Very low	Angola, Benin, Burundi, Chad, Central Africa Republic, Congo, Djibouti, DRC, Gambia, Equatorial Guinea, Eritrea, Gabon, Guinea, Guinea Bissau, Lesotho, Liberia, Niger, Sierra Leone, Togo, Somalia, South Sudan
Low	Botswana, Burkina Faso, Cameroon, Côte d'Ivoire, Eswatini, Ghana, Madagascar, Namibia, Malawi, Mali, Mozambique, Rwanda, Senegal, Sudan, Tanzania, Uganda, Zambia, Zimbabwe
Medium	Ethiopia, Kenya, Nigeria
High	South Africa
Very high	None

[a]Countries not listed due to lack of sufficient data: Cape Verde, Comoros, Mauritania, Sao Tome and Principe and Seychelles

universities of agriculture within SSA have training and research programmes in crops and to some extent livestock, but only a few have such programmes on forestry and aquaculture.

Because of the large contribution of crop agriculture to feeding SSA countries and the myriad of production challenges facing the sector, countries have deliberately and consistently prioritized the sector. Livestock has also consistently been the second priority for most countries, with forestry and aquaculture receiving less focus.

A link is also apparent between capacities for crops and forestry and between livestock and fisheries for some of the countries. For example, in DRC and Ethiopia, resources appear to be somewhat shared between the two pairs of sectors, such that high capacity in crops flows to and contributes to that of forestry, while livestock corresponds to that of aquaculture. However, this desirable practice was limited in scope across countries—even in those countries which seem to have recognized its merit.

8.3   Enabling Environment for Biotechnology Applications Across Sectors of Agriculture

While application of biotechnology in agriculture has demonstrable potential, this is only possible if certain enablers are in place. Human and infrastructural capacity are key enablers that countries need but are struggling to have. Countries also need other enablers. These additional enablers are discussed here under three main headings: public awareness and political support; policy environment (including biosafety frameworks), defined as the expressed and documented intent and courses of action aimed at promoting technology development, such as commitments and actions related to addressing capacity needs and setting up of institutional mechanisms that facilitate safe use of technologies (without becoming a hindrance to R&D processes); and R&D resourcing (both public and private investments).

Public Awareness and Political Support Due to controversy surrounding genetically modified organisms (GMOs) in agriculture across SSA, there is more public scrutiny of the application of this technology than many others. Initially focused primarily on GMOs, the media is awash with articles and stories demonstrating, on the one hand, the usefulness of biotechnology to farmers and, on the other, scepticism and outright opposition. Unfortunately, these misperceptions and misrepresentations are often extended to any conversation about agricultural biotechnology, and this is creating an environment that is increasingly characterized as overwhelmingly anti anything biotechnology, not just GMO.

With the exception of South Africa, there are no comprehensive national surveys assessing the public understanding, perception and acceptance of biotechnology in Africa. The South African study (Gastrow et al. 2016) reveals a very high level of ignorance about biotechnology among the general population and favourable support for biotechnology among better informed respondents. Despite the perception that the public is aware about biotechnology and what it can do or not do, much of the paranoia can be attributed to misperception due to lack of understanding. Thus, it is safe to say that progress made by many SSA countries in establishing programmes of research and applications of biotechnology has not necessarily been a reflection of public consensus. It is, by and large, a reflection of political commitment, informed and driven by the technicians and practitioners. Public awareness remains a gap, even in countries like South Africa that rank high in applications of, and capacities for, agricultural biotechnology.

The ongoing coronavirus pandemic and the roll-out of mass vaccination campaigns have provided some insights into the acceptance of advanced biotechnology. The Oxford-Astra Zeneca vaccine, for example, is a viral vector vaccine that consists of a Chimpanzee adenovirus that has been genetically modified so it cannot replicate or cause disease in the patients to which it is administered and also so that it contains the gene that codes for the spike protein on the surface of severe acute respiratory syndrome coronavirus 2 (SARS-CoV-2), the coronavirus that causes Covid-19. Box 8.2 presents the case of vaccine hesitancy in Africa and elsewhere as an example of technology acceptance challenge even in human health in the face of a major pandemic.

Box 8.2 Vaccine Hesitancy in Africa and Beyond

Vaccine hesitancy—defined as delay in acceptance or refusal of vaccination despite availability of vaccination services—is an issue around the globe. Even before the ongoing Covid-19 pandemic, the World Health Organization (WHO) identified vaccine hesitancy as being one of the top ten global health threats in 2019 (WHO, no date). Estimates for the proportion of the population that needs to be vaccinated to achieve herd immunity for Covid-19 vary but could be as high as 80–95%, so vaccine hesitancy is a real challenge.

A recent global review published in *Nature* and involving 19 countries, including Nigeria and South Africa, reported that, overall, 71.5% of all respondents stated they were 'very likely' or 'somewhat likely' to take a Covid-19 vaccine. Stated acceptance rates (which may vary from actual acceptance) varied between countries: it was highest in China (close to 90%) and lowest in Russia (55%); the acceptance rate for Nigeria was 65% and for South Africa almost 80%—for comparison, the rate for the USA was 75%. Middle-income countries, including South Africa, had relatively high acceptance rates. Older respondents were more likely to accept the vaccine than younger respondents. Respondents with higher incomes and higher education levels were also more likely to accept the vaccine (Lazarus et al. 2021). Africa has a young population that is mostly poor and basically educated, so these results are a cause for concern.

A survey carried out in Cameroon reported a low Covid-19 vaccine acceptance rate of just 15% in a cohort of young adults compared to 56% in DRC. The latter included low acceptance rates even among health-care workers. In contrast, a survey in Rwanda reported 86% of respondents willing to accept a vaccine (Nachega et al. 2021).

One of the factors associated with higher vaccine acceptance rates, as reported in the *Nature* review, was trust in the government. In this context

---

(continued)

Box 8.2 (continued)

the lack of, or failure to demonstrate, political will by some African heads of state is worrying. John Magufuli, who was then president of Tanzania, was described as 'Africa's most prominent Covid denier'. He stated in January 2021 that 'there is no coronavirus in Tanzania', scoffed at the use of facemasks, criticized neighbouring countries' lockdowns, and rejected vaccines. He died in March 2021, aged 61, and paradoxically was said to have died from Covid-19 complications. Meanwhile, in Madagascar the president has promoted an herbal tea, developed by the Madagascar Institute of Applied Research, which he has publicly claimed prevents and cures Covid-19.

In contrast, leadership on pandemic response from the African Union, chaired by South Africa during 2020, has been regarded as generally positive and a factor responsible for the relatively low death rate observed on the continent, although other factors, such as the youthful population, the low rate of obesity and the hot climatic conditions are also considered to be significant.

Clearly, political commitment and public awareness will remain important factors in determining acceptance and wide use of new technologies. Globally, human health is one domain where modern biotechnology has been accepted more readily.

Policy and Biosafety Frameworks Generally, key elements of policy and biosafety were driven by actions taken for and by the crops sector: progress specific to the other three sectors has tended to build on foundations built for the crop sector. To the extent that crops remain the highest priority agricultural sector for most countries, and because early progress made by the international community in developing critical enabling environment frameworks was crop-centric (e.g. key underpinnings for developing biosafety laws), this trend was expected.

The four sectors of agriculture—crops, livestock, forestry and aquaculture—are almost invariably administered under different government ministries. Even where consolidation has been done to try to enhance administrative efficiency, it is rare to find all four sectors under one government ministry. This dispersed administrative structure for agriculture and the lack of efficient coordination mechanism hamper technology applications. Moreover, placing two or more of these sectors as departments in one ministry does not seem to necessarily address the major coordination challenges that undermine efficiency in programming.

With regard to creating an enabling environment for the applications of biotechnology, a key missed opportunity is taking full advantage of the similar nature of most applicable tools for biotechnology and developing national R&D infrastructures and associated institutional arrangements that facilitate sharing of facilities and human resources across sectors. If done, this could enhance efficiency in resource use and effectiveness in delivery of biotech outcomes. For the most part, the limited capacity, such as labs and personnel, is designated to serve specific sectors, even in countries that are significantly resource-constrained. There was very limited evidence for, and cases of, deliberate efforts aimed at coordinated cross-sectoral action in the areas of biotechnology. In the absence of this, opportunities for synergies across sectors are being missed owing to lengthy and messy bureaucratic procedures required to share physical and human resources. For most countries, sharing of these resources across ministries is next to impossible. Even canvassing for funding is done separately for each sector, and the amount allocated to the hosting ministries is not necessarily consistent with the needs of, and state of development in, the respective sector. One of the best examples of integrated and coordinated cross-sectoral collaboration for biotechnology R&D is that of the Forestry and Agricultural Biotechnology Institute at the University of Pretoria, South Africa (Chap. 6, Box 6.1), where human resources and high-end facilities are shared across forestry and crop sectors.

Public and Private Investments While the formulation of policy and establishment of biosafety frameworks are principally a function of the political will of the country—in turn driven by an understanding and acceptance of the potential of biotechnology, not necessarily resource endowment—a major aspect of the enabling environment that seems to challenge the majority of SSA countries, even countries with strong political will, is resourcing of biotechnology programmes. This includes investments in capital items, such as labs and equipment, human resources and operations. Although precise values of agricultural biotechnology spending are difficult to obtain, estimates focusing only on crops and livestock obtained from the ASTI database show that SSA countries invest very limited amounts on agricultural R&D generally, and agricultural biotechnology in particular (Fig. 8.4). The total agricultural R&D spending takes the same pattern as in crops and livestock; South Africa, Kenya and Nigeria are consistently among the top in terms of agricultural R&D spending.

Private investments in agricultural R&D in SSA are mainly directed towards high-value crops and non-traditional highly commercial products such as cut flowers. Fruit and vegetable exports, especially from East Africa, are also experiencing relatively high growth. A recent development is the proliferation of private agribusiness investment funds targeting African agriculture. Similar to the case of land purchases, most of the funds have recently been set up and are still in the fundraising stage of their development. There have also been encouraging developments at both continental and national levels in efforts aimed at stimulating additional private sector agribusiness investment. In addition, although progress is slow since the Maputo Declaration in 2003, the position as of 2015, that is the lead up to Malabo Declaration, indicated that some countries had taken steps to honour their commitments to increasing investments in agriculture and a number of countries have taken a proactive role in attracting private sector agribusiness investments by offering various incentives such as tax holidays within the first few years of an agribusiness establishment (e.g. Nigeria) and zero duty on agricultural equipment (e.g. Ghana and Nigeria). There are also a number of national initiatives across the continent for advancing agribusinesses. This pattern of growing private sector interest in, and content of, agricultural value chains will continue. What is

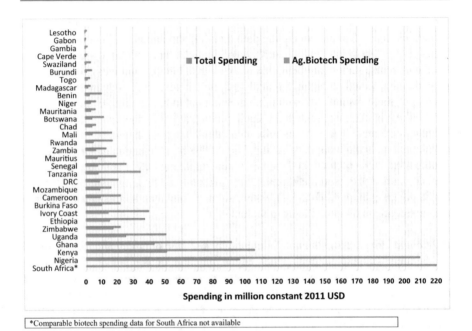

Fig. 8.4 Total ARD and biotech spending (2011 USD in millions) in SSA (2014)* (*Comparable biotech spending data for South Africa not available)

important, however, and must be emphasized, is not the mere number of initiatives, but their effectiveness in mobilizing private sector investment and engaging agribusiness in research and commercialization of biotech products.

Other than South Africa, whose comparable data were not available for this analysis, the top countries in total agricultural biotech spending were Nigeria (USD96.4 million), Kenya (50.8 m), Ghana (42.9 m) and Uganda (25.2 m). The figures show that even these leading countries spend only modest amounts on biotech. Other countries spending more than USD10 million on crop and livestock biotech are Burkina Faso, Côte d'Ivoire, Ethiopia and Zimbabwe. An estimated 40% of SSA countries spend less than USD5 million on crop and livestock biotech a year. Mobilization of resources from domestic and external sources for agricultural biotech is clearly a major area that governments need to look at. In the meantime, given the high cost of biotech R&D, available investments need to be used in a much more coordinated manner to achieve efficiencies from scale and complementarity— and hence the need for cross-sector coordination. More intentionality is clearly required to ensure that programming in these sectors is better coordinated, and they work in synergy to optimally use the limited human, infrastructural and financial resources available in a country.

Despite the clear dominance of the public sector both in the financing and implementation of agricultural research, the unstable funding of agricultural R&D to date suggests that other avenues should be explored. Universities in Africa are an underutilized resource that could greatly increase research output with just slight increases in targeted funding to them.

Collaboration and Networking The entry of African countries into biotechnology has been stimulated by many interrelated factors, particularly the cumulative nature of the advances in biotechnology. In addition, the pace of SSA biotechnological advancement has benefited from regional, sub-regional organizations and networks that have been rightfully credited with the development of broader agricultural R&D capacity in SSA (Chambers et al. 2014) and that have also contributed in significant ways to many aspects of enabling environments. These include the biotechnology support programmes and initiatives driven by the CGIAR whose centres have, for four decades, worked collaboratively with many SSA countries on biotechnology research and applications in different sectors (Kiome 2015) although the countries hosting the centres for a relatively larger share of this.

In the livestock sector, ILRI working in partnership with national and other international partners has made strides in developing genetically engineered vaccines, while in forestry ICRAF has provided a lot of support in capacity development as well as research and application of low- to medium-level forestry biotechnologies. It is, however, the CGIAR centres in the crop sector that have made the greatest contributions, again related to the priority given to crop agriculture in the SSA region and globally. Several CGIAR centres, including ICRISAT, CIMMYT, CIP and IITA, have contributed substantially in the research and application of medium- to high-level biotechnology in maize, potatoes, cassava and sorghum among other crops. Continental research organizations such as AGRA and AATF have also played a part in research and development as well as application of biotechnology in the crop sector (Kiome 2015; Nang'ayo 2006).

Another critical set of institutions whose strategic approaches emphasize the centrality of networking and collaboration among countries are the sub-regional organizations (SROs) and the apex agricultural R&D organization, the Forum for Agricultural Research in Africa (FARA). The SROs in Africa were established in the early 1990s based on the realization by countries of the potential gains from regional cooperation in addressing common constraints to agricultural production. The aim was to ensure effective co-coordination and develop collaboration in agricultural research and development in the respective sub-regions, creating synergies to reduce duplication and mobilizing resources to catalyse technology generation, dissemination and adoption through collective research, training and capacity building while also providing fora for regional dialogue on agricultural policy bottlenecks. The SROs were designed to complement activities of the NARS. Other functions envisaged for the SROs included knowledge management and sharing, advocacy for support of agricultural research and fostering partnerships with other research institutions at regional, continental and international levels. The SROs have evolved over time, and their roles have morphed: at various points they have managed competitive agricultural R&D grants from multilateral donors. The current SROs in SSA are the Association for Strengthening Agricultural Research in Eastern and Central Africa (ASARECA), which provides technical guidance to NARS in its

11 member states; the West and Central African Council for Agricultural Research and Development (WECARD/CORAF) (23 countries); and the Centre for Coordination of Agricultural Research and Development for Southern Africa (CCARDESA) (15 countries). FARA was conceived to be an umbrella organization bringing together major stakeholders in agricultural research for development in Africa. It was tasked to facilitate research by the other constituents including SROs, NARS and their partners at national and sub-regional levels.

Other significant networks and initiatives that contribute to application, capacity and creation of enabling environment for biotechnology in SSA include the AU/NEPAD Africa Biosafety Network of Expertise which consolidates capacity, networks and information resources for regulators in order to build functional biosafety systems in Africa; the UNEP-GEF capacity building programme on biosafety that has worked to strengthen African countries' capacity development of biosafety regulatory and national frameworks in compliance with the Cartagena Protocol on Biosafety (Makinde et al. 2009); and the AU/NEPAD Biosciences Initiative which has developed Biosciences Centres of Excellence (as exemplified by BecA—see Chap. 2) in different sub-regions to strengthen capacities for biosciences in these sub-regions. The regional and sub-regional agricultural research organizations and RECs such as FARA, ASARECA, CORAF/WECARD and COMESA have all contributed in different ways to improve the overall enabling environment through human and institutional capacity strengthening for R&D, including the development of policies and biosafety frameworks (Morris 2011; FARA 2011).

There was a close relationship among the thematic areas of the analysis: capacities, enabling environment and application. Higher capacities corresponded with higher levels of application and enabling policy environment. A relationship cycle can explain this observation—a stronger enabling policy environment promotes higher capacity and hence enables technology development and application. On the other hand, a country cannot regulate nothing—robust biotechnology research and application would require and hence catalyse the development of regulation, policy and laws, for example. Policies and legislations on biotechnology in a country with no research on, and/or application of, biotechnology are meaningless unless they are part of a plan. However, overall, having a critical mass of requisite human capacity emerged as the critical starting point.

Table 8.2 summarizes the status of countries across the four sectors on the basis of enabling environment (see also Ochieng and Ananga 2019). From the available information, only South Africa was considered to have a 'very strong' enabling environment for agriculture, but this excludes the aquaculture sector. Five countries, Ethiopia, Ghana, Kenya, Nigeria and Sudan, were considered to have a 'strong' enabling environment, while for a further eight countries (Botswana, Malawi, Mali, Namibia, Tanzania, Uganda, Zambia and Zimbabwe), this was rated 'medium'. This leaves 29 countries considered to have 'weak' or 'very weak' enabling environments for agricultural biotechnologies.

Table 8.2 Classification of countries across sectors on basis of enabling environment

Enabling environment across sectors	Countries
Very weak	Angola, Benin, Burundi, Chad, CAR, Congo, Djibouti, DRC, Eritrea, Gambia, Equatorial Guinea, Eswatini, Gabon, Guinea, Guinea Bissau, Lesotho, Liberia, Niger, Togo, Sierra Leone, Somalia, South Sudan
Weak	Burkina Faso, Cameroon, Côte d'Ivoire, Madagascar, Mozambique, Rwanda, Senegal
Medium	Botswana, Malawi, Mali, Namibia, Tanzania, Uganda, Zambia, Zimbabwe
Strong	Ethiopia, Ghana, Kenya, Nigeria, Sudan
Very strong	South Africa (except aquaculture)

see also Ochieng and Ananga (2019)

8.4   Applications of Biotechnologies Across Sectors of Agriculture

Figure 8.5 is a graphical summary of the classification of countries on the basis of application of biotechnology, while Table 8.3 is an attempt to categorize countries (from very low to very high) by level of biotechnology applications across all the four sectors—crops, livestock, forestry and aquaculture. Of note is the fact that there were major differences among sectors both within and across countries in the extent of applications of biotechnologies, with crops consistently having higher levels of applications, followed by livestock, forestry and aquaculture, in that order.

Across sectors, no country out of the 43 countries studied is classifiable under 'very high' use, but one country (South Africa) falls in the 'very high' use category for crops. Crops also lead in terms of number of countries (9) in the 'high' and 'medium' (17) use categories. Livestock does not feature in the 'very high' use category but is second to crops both in the 'medium' and 'high' use categories, with South Africa and Kenya being the only countries in the 'high' use category. Except for South Africa classified in the 'high' use category, biotechnology applications in forestry in SSA fall predominantly in the 'very low' to 'medium' categories. All aquaculture applications are in the 'very low' to 'medium' categories.

Across the four sectors, Kenya and South Africa are 'high' users, while 'medium' users are Madagascar, Malawi, Nigeria, Sierra Leone, Sudan, Tanzania, Uganda and Zimbabwe. The rest of the countries are in the 'very low' (21) and 'low' use (12) categories.

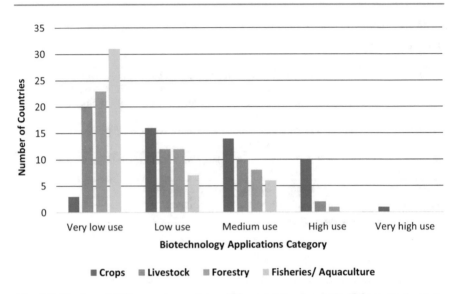

Fig. 8.5 Number of SSA countries applying different levels of agricultural biotechnologies by sector

Table 8.3 Classification of countries based on biotechnology applications across the four sectors

Applications category	Countries[a]
Very low use	Angola, Benin, Burundi, CAR, Chad, Congo, DRC, Eritrea, Equatorial Guinea, Eswatini, Gambia, Guinea, Guinea Bissau, Côte d'Ivoire, Lesotho, Liberia, Mali, Niger, Rwanda, Somalia, Togo
Low use	Botswana, Burkina Faso, Cameroon, Djibouti, Ethiopia, Gabon, Ghana, Mozambique, Namibia, Senegal, South Sudan, Zambia
Medium use	Madagascar, Malawi, Nigeria, Sierra Leone, Sudan, Tanzania, Uganda, Zimbabwe
High use	Kenya, South Africa
Very high use	None

[a]Countries not listed due to lack of sufficient data: Cape Verde, Comoros, Mauritania, Mauritius, Sao Tome and Principe and Seychelles

8.5 Overall Status of Agricultural Biotechnology Across Sub-Saharan Africa Countries

Table 8.4 shows the overall relative status of agricultural biotechnology in SSA countries. This table was generated by combining the results for capacity, enabling environment and application of agricultural biotechnologies across all four agricultural sectors. For this, scores were allocated to each country corresponding to the band in which they fell with respect to capacity, enabling environment and

Table 8.4 Overall status of biotechnology in SSA countries across sectors and factors[a]

Overall status of biotechnology	Agriculture contributes more than 20% to GDP	Agriculture contributes less than 20% to GDP
Weak	Benin, Burkina Faso[b], Burundi, CAR, Chad, Côte d'Ivoire, Gambia, Guinea, Guinea Bissau, Liberia, Mali, Niger, Rwanda, Sierra Leone, Somalia, Togo	Angola, Botswana, Cameroon, Congo, Djibouti, DRC, Eritrea, Equatorial Guinea, Gabon, Ghana, Lesotho Senegal, South Sudan, Eswatini[b]
Medium	Ethiopia[b], Madagascar, Malawi[b], Mozambique, Nigeria[b], Tanzania, Uganda	Namibia, Sudan, Zambia, Zimbabwe
Strong	Kenya[b]	South Africa[b]

[a]Capacity, enabling environment and application
[b]Countries which have approved commercial release of GM crops

applications. The scores ranged from 1 to 5, where 1 was very low/very weak and 5 very high/very strong. The scores for each country from these three tables were added together. Based on these total scores, countries were allocated to one of three categories—weak, medium or strong—for overall biotechnology status.

Only Kenya and South Africa are considered to have an overall agricultural biotechnology status of 'strong'. Eleven countries fall in the 'medium' band and the majority (30) are categorised as 'weak'.

Not surprisingly, five of the seven countries that have permitted commercial release of GM crop varieties are in the 'strong' or 'medium' category. Burkina Faso is an outlier as it has previously permitted commercial release of GM cotton, but this was later suspended (see case study Sect. 4.5.2 in Chap. 4). This is, perhaps, a cautionary tale of the potential danger of allowing field application to be out of step with the status of capacity and the enabling environment. Eswatini, a relatively small country, is also an outlier. The country's agriculture is, however, closely linked to that of South Africa, and it has a highly productive, industrialized agriculture sector growing high value crops, especially sugar, forestry and citrus, with high levels of investment and irrigation.

Sixteen countries where the national economy is highly dependent on agriculture (contributing more than 20% to GDP) fall into the 'weak' category for overall biotechnology status (Benin, Burkina Faso, Burundi, CAR, Chad, Côte d'Ivoire, Gambia, Guinea, Guinea Bissau, Liberia, Mali, Niger, Rwanda, Sierra Leone, Somalia and Togo). Without the benefits delivered by modern and advanced biotechnologies, it is hard to see how these countries will achieve the growth rates in agricultural productivity needed to meet the Africa Union's 6% annual target or to achieve food security and provide sustainable livelihoods for their rapidly growing populations. Significant investment and support will be needed to enhance the biotechnology status of these countries to a point where they can apply modern and advanced biotechnologies widely across sectors. With many of these countries being located in the West African region, there is a strong argument to be made for regional programmes, including access to shared, regional high-level biosciences facilities and expertise, similar to that provided by the ILRI-BecA Hub which, to date, has primarily served eastern and central Africa.

In contrast, this analysis identifies eight countries (Ethiopia, Madagascar, Malawi, Mozambique, Nigeria, Tanzania, Uganda and Kenya) whose economies are highly dependent on agriculture and whose overall biotechnology status is 'medium' or 'strong'. These countries appear to have much higher potential to reap the benefits of modern and advanced biotechnologies, enabling their agriculture sectors' productivity to grow and to cope better with adverse impacts of climate change. Most are also located so they can more readily take advantage of the facilities and expertise provided by the biosciences facilities at the ILRI-BecA Hub.

8.6   Gaps and Opportunities for Agricultural Biotechnology

Besides classical varietal development and tissue culture in crops, phenotypic and molecular characterization mostly in crops and livestock and artificial insemination in livestock, the use of modern advanced technologies remains very limited in most SSA countries, largely confined to research projects. However, use of molecular and genomic technologies, while still low, is increasing rapidly—especially in research. While technical limitations exist, the major constraints to the application of most of these technologies in SSA relate to human, infrastructural and institutional capacities and various dimensions of the enabling environment—policy and legislation, ethical concerns and financial investments.

Building on the analysis presented in the preceding chapters, in this chapter we identify ten interrelated critical gaps and their corresponding opportunity areas for action to enhance research and applications of biotechnologies in SSA agriculture. The gaps and opportunities are human capacity; research infrastructure (facilities and equipment); financial resources; policies and biosafety frameworks; taking research products in to use; public-private partnerships; intellectual property protection; administrative structure and inter-sectoral coordination; and awareness and public participation. In addition to these nine areas, we highlight, as strategic intervention area 10, the relatively low priority consistently given to forestry and aquaculture across most SSA countries—even where the relative contributions of these sectors are high—as an area that needs attention.

1. *Human capacity.* Most SSA countries lack a critical mass of scientists in areas relevant for agricultural biotechnology. Even countries that rank as relatively 'high' in capacities often lack critical mass in the more advanced areas of modern biotechnology, such as molecular biology, genomics and bioinformatics. Indeed, NARS in many SSA countries have research programmes that are often limited in scope and dependent on a handful of scientists. Compounded by infrastructural and financial constraints (see 2 and 3 below), many such programmes have limited national capacities to implement promising initiatives beyond the pilot scale. It would be beneficial, especially for the weakest countries, to establish and/or strengthen biotechnology multi-sectoral R&D networks at national,

sub-regional and continental level in order to create synergy, facilitate horizontal and transboundary transfer of technologies, enhance better utilization of resources, avoid duplication and upscale best practices and technologies. More-over, there are also gaps in the capacities of the regional and sub-regional research organizations (SROs) to assist NARS in the establishment and development of partnerships with high-level biotechnology research facilities around the world and also international training opportunities that can encourage rapid development of biotechnology in Africa. The strategies of FARA and SROs have historically included capacity development of NARS, focusing mostly on opportunistic training courses and time-bound projects driven by donor funding. While these have delivered some capacity outcomes, claims made of significant impacts have been exaggerated: these on-and-off interventions have not left a lasting mark on SSA agricultural R&D landscape. A new approach is needed, one that focuses on building capacities of institutions in sustainable ways. This needs to include 'midwifing' the establishment of innovative institutional arrangements among NARS and between NARS and other regional, continental and international institutions which are at the cutting edge of science in specific priority areas, as a basis for ongoing exposure and retooling of current researchers, research and hardware calibration/maintenance technicians, as well as platforms for training next-generation researchers and research leaders.

2. *Research & development infrastructure.* The acquisition and maintenance of the expensive infrastructure needed for high-tech applications, such as genetic engineering, remain a challenge for most countries. Many countries are so constrained by research infrastructure that they cannot undertake even the most basic biosciences research. In other cases, equipment acquired through projects only functions during the life of these projects and thereafter is rendered useless because it cannot be maintained due to budget constraints. The lack of local engineers and technicians trained to service the fast-evolving and sophisticated equipment presents another challenge, as does inadequate power supplies and frequent power outages. The latter presents a special challenge not just to running of equipment in the labs, but also to the development of reliable cold chains, e.g. for AI and vaccine field delivery. These infrastructural capacity challenges for biotechnology applications have formed the arguments for the establishment of regional shared biotechnology platforms, such as the BecA-ILRI Hub (see Box 2.3, Chap. 2). Experiences to date suggest that this is one of the most promising opportunities for SSA as a means of supporting access to facilities which NARS need for their current research while also informing strategies for stepwise development of national infrastructural capacities. While the establishment of BecA was principally driven by an international centre, ILRI, and was designed to serve many countries, the concept of shared labs could be pursued through bilateral or multilateral agreements between or among a cluster of neighbouring countries. The East Africa Agricultural Productivity Programme (EAAPP) funded by the World Bank started the establishment of commodity-specific facilities in different countries across the region—e.g. dairy in Kenya and cassava in Uganda. The idea was that these facilities would be further developed and

maintained by the host countries and continue to be accessed by regional members of the 'network'. Unfortunately, the programme design did not pay adequate attention to sustainability. Use of these facilities has principally remained domestic and under-utilized, and their maintenance has become a challenge for the perpetually under-funded NARS institutions. Sharing of facilities is justified by the fact that many countries are not in a position to have their own fully equipped facilities that cover the spectrum of technologies relevant for them; most of these facilities are so expensive that the only way to get decent returns to investments is through economies of scale—to ensure that they are used intensively (many countries are unlikely, e.g. to use the full capacity of a sequencer); and much of the equipment is generic and applicable across multiple commodities (plants, animals, microbes) or disciplines (animal genetics, diagnostics, nutrition). Given the current limited numbers of active biotech facilities, which are also geographically thinly spread across regions and countries, innovative institutional arrangements are clearly needed to address the current gap in equipment maintenance and repair. An example could be maintenance contracts between a service provider and a consortium of labs across countries, possibly linked to shared biotech platforms. This would ensure that there is high enough demand to spread the associated overheads. A challenge to FARA and SROs is how to catalyse the establishment of innovative institutional arrangements (see also 1 above) of this or similar types. This is a perfect fit for the mandates of these institutions. Indeed, this is more relevant than managing the modest pass-through grants to individual countries that have, to date, not made any significant difference.

3. *Financial resources for R&D.* A major challenge for public agricultural biotechnology R&D in Africa remains how to mobilize investment capital, beyond what is needed for infrastructure, to initiate or sustain research and facilitate the process of taking findings to commercial use (see 5 below). Although there has been some growth in the level of funding to agricultural biotechnology R&D in some countries, the level of financing is still extremely low and does not allow countries to engage effectively in cutting-edge biotechnology research. Most of the current biotechnology R&D programmes are donor funded with very limited national budget content; in most countries the so-called national allocations only cover salaries of the limited number of staff. Although precise estimates of the value of agricultural biotechnology spending are difficult to obtain, estimates made on the basis of 2014 figures (focusing only on crops and livestock) obtained from the ASTI database show that most countries invest very limited amounts on agricultural biotechnology (see Sect. 3.2.3 in Chap. 3). Only four countries (South Africa is also in this category but not included in the ASTI database) had spending levels above USD20 million: Nigeria was the largest spender (USD96.4 million), followed by Kenya (50.8 m), Ghana (42.9 m) and Uganda (25.2 m). Among the other countries spending more than USD10 million on crop and livestock biotech are Burkina Faso, Côte d'Ivoire, Ethiopia and Zimbabwe. About 40% of SSA countries spend less than USD5 million on both crop and livestock biotech research a year.

4. *Biotechnology policies and biosafety frameworks.* The extent to which biotechnology has contributed to agricultural productivity in various countries is closely linked with, and has been dictated by, the policy/political landscape and the nature of legislation enacted to govern the technology process. The lack of biotechnology policies and biosafety legislation, and absence of biosafety procedures in several countries, continues to be a significant impediment and discouragement to biotech research programming in most SSA countries. They are unable to undertake high-level biotech R&D because they are not able to obtain approvals from regulatory authorities, or because processes for application are opaque and tedious, and generally the institutional landscape does not encourage R&D with significant biotech content. This is not only an impediment for the emergence of a robust and passionate community of R&D practitioners but is also strongly correlated with private sector investments in biotech R&D. The relationship among application, capacity and enabling environment is complex, each in some way acting as driver for the others, with mutually re-enforcing effects. The enabling environment and capacity for biotech applications are especially closely linked. In the first place, the development of relevant policies and regulatory frameworks requires competent inputs by technical experts—the 'biotechnologists'. Looked at this way, absence of enabling biotechnology and biosafety laws can be attributed to the lack of capacity for biotechnology and biosafety policy-making. Examination of the trajectories that most 'high' biotech status countries have gone through, in terms of applications, capacities and enabling environment, leads to the conclusion that the enabling policy and regulatory environment are, in early phases, principally driven and shaped by the demand (applications) side—and not just the existence of capacity. That is, it starts and evolves at a pace that reflects the level of vibrancy on the applications side. It is from that angle that capacity for applications comes into play. In other words, governments could develop the required frameworks, but they will serve very little purpose and will not evolve if they are not being subjected to real tests through active applications. Indeed, the fact that African countries face challenges in the implementation of the Cartagena Protocol on Biosafety is directly related to the generally weak capacities to enact, implement, monitor and enforce national biosafety laws. Moreover, while many African countries have ratified and acceded to international instruments/regimes related to biotechnology, domesticating these instruments, coordinating complex institutional arrangements and harmonizing overlapping and conflicting mandates still remain a formidable challenge and a persistent gap. Clearly, capacities, enabling environment and applications are so intricately linked that only an intentional multi-pronged approach can overcome the apparent chicken-and-egg situation most countries find themselves in.

5. *Research into use.* The process by which biotech research translates to commercial applications in the field requires early engagement of industry including farmers and other private sector value chain actors. But agricultural biotech research in almost all SSA countries is still primarily driven by NARI and university scientists, who have limited knowledge on how research products can be commercialized. At the same time, public extension services are generally weak. There is clear need for countries to develop capacities and multistakeholder

mechanisms for technologies to be put into use through private sector-driven process. Biotech incubators are widely recognized globally as a way to support nascent business development and to bridge the gap from research to commercialization, and this has long been advocated by the African Union (Juma and Serageldin, 2007). However, African governments have not embraced this, and there are only very limited examples of bio-incubator initiatives—all at very early stages—on the continent, e.g. BioPark Mauritius (Box 8.3). Other similar budding examples with potential include:

- The Forestry and Agricultural Biotechnology Institute (FABI) at the University of Pretoria (http://www.fabinet.up.ac.za/) which holds promise as a destination for the development of agrochemicals
- The International Centre for Genetic Engineering and Biotechnology (ICGEB) (http://www.icgeb.org/home-ct.html) in Cape Town, South Africa, which has the capability to advance the understanding of human disease from various perspectives
- Biovac in South Africa (http://www.biovac.co.za/) set to develop and make vaccines for the African market
- The Uganda Biosciences Information Center (UBIC) (http://www.ugandabic. org/n/) which has been set up to disseminate information on biotechnological innovations and advances in agricultural research in Uganda

---

Box 8.3 BioPark Mauritius: Helping Invigorate Biotechnology Research in Africa

An example of an initiative that helps invigorate biotechnology in Africa is BioPark Mauritius (http://www.bioparkmauritius.mu/). One of the first multidisciplinary hubs in Africa, it provides a dedicated space for research and development in the biomedical sector.

BioPark Mauritius comprises state-of-the-art infrastructure and facilities covering an area of over 100,000 square metres. The initiative receives a share of the less than 1% of the annual GDP (USD11.5 billion in 2015) allocated to the R&D sector by the Mauritian government.

Its main goal is to attract multidisciplinary research in the life sciences including microbiology, chemistry, toxicology, pharmacology and epidemiology. As of March 2016, five biotech companies and two contract research organizations had established their operations at BioPark Mauritius. One of the companies is focused on the invention of environmentally friendly medical consumable supplies such as syringes, gloves and blades. It has recently been reported that 5% of all catheters sold worldwide are produced by leading European medical device companies that have set up shop in BioPark Mauritius.

(continued)

Box 8.3 (continued)

While this initiative is not focused on agricultural biotechnology, it points to the successful trajectory that the life sciences research could take in Africa and represents an opportunity that SSA governments need to look into seriously as part of a holistic strategy to better harness the potential of biotechnology.

Source: Compiled by authors

6. *Role of public-private partnerships* Although public-private partnerships (PPPs) have been recognized as one of the ways of driving the conversion of biotech research into practical use, and despite the fact that there are a number of PPPs operating in some countries, there remains major gaps in operationalizing the concept and developing functional partnerships at scale. PPPs are a way of ensuring that private research in certain biotech areas, e.g. GM crops, is combined with local knowledge of varieties and cropping conditions that resides in public research organizations in order to develop GM crops suitable to African conditions. A major gap in PPPs is the issue of proprietary rights, especially patents, intellectual property rights and sharing of benefits accruing from joint biotechnology research and development activities (see 7 below, *Intellectual property protection*).

7. *Intellectual property protection.* Recognition of the role of intellectual property protection and its impact on the acquisition, development and diffusion of biotechnology remains a big gap in most countries. Institutions for administering industrial property rights, particularly patents, are still in their infancy. While a good number of SSA countries have established patent offices, the utility of these as sources of scientific and technological information has not been adequately exploited. There is also a growing debate on the impact of intellectual property protection on the transfer of modern biotechnology to African countries. Concerns on this issue are largely based on the view that intellectual property protection is a barrier to transfer of technology.

8. *Administrative structure and coordination.* In SSA countries, the four agricultural sectors—crops, livestock, forestry and aquaculture—are not necessarily under the same government ministry or department. Indeed, in some countries all of these sectors are in different ministries, although finding livestock and crops under the same ministry is more common. The administrative separation of these sectors, combined with poor cross-sectoral coordination that is characteristic of most government functions in SSA, is inimical to efficiency in the development and use of technologies. Specifically, it is responsible for the poor leveraging of gains made and capacities available primarily in the crops sector to leap-frog progress in the other sectors. It limits consolidation and exploitation of synergies across sectors owing to bureaucratic procedure required to share physical and human resources, such as labs and personnel. For most countries, sharing of these resources across ministries is impossible. Even mobilization of resources is done separately by sector, and the amounts allocated to the hosting ministry do not

always reflect the sector needs. For efficiency in resources use and effectiveness in delivery, there is a major need for better coordination among these sectors with a specific view to ensuring that the development of each sector reflects its relative importance and leverages on the other sectors as much as is possible.

9. *Public awareness and participation.* There are major gaps in public awareness and clear understanding of the science and the potential promise and usefulness of the whole spectrum of biotechnology in African agriculture. There are also knowledge gaps. Indeed, misinformation on risks and perceptions of risks remains one of the key factors that have hindered the adoption of biotechnology in the continent, especially GMOs. There is also confounding with outright pushbacks against the involvement of multinational companies, such as Monsanto/Bayer, that are unrelated to safety but speaking to issues around 'decolonizing develop-ment', national sovereignty, unfair profiteering, farmers' rights and social justice. Unfortunately, the success of the anti-GMO campaigners, and part of the strategy by this group, seems to depend on putting all the issues in single baskets and that way creating enough confusion to deflect attention from available scientific evidence base. The campaign against the Water Efficient Maize in Africa (WEMA) Project[1] is a good example. Although there have been successes in public awareness creation, there are still gaps in policy support, political commit-ment and acceptance of genetic engineering technologies, and this continues to hinder the adoption of certain biotechnologies in agriculture. While the lack of capacity seems to be an issue in several countries, some appear to be adopting a wait-and-see attitude towards agricultural biotechnology, and this is stalling investments and slowing down progress. Strategies are needed to help countries deal with the array of challenges related to awareness, acceptance and public participation. The strategies should reflect the situation in each individual country. Practical applications of biosciences in agriculture should be taught in primary, secondary and high school curricula to expose the next generation to the opportunities and risks early enough so that they grow up to become better informed leaders, policy-makers and consumers able to make objective choices.

10. *Limited attention to forestry and aquaculture.* In all SSA countries, the crop sector has been at the forefront in the development and use of biotechnology. The sector leads in applications, capacities and enabling environment. Indeed, policies and regulatory frameworks relevant for agricultural biotech in SSA have, almost without exception, been developed in response to, or in anticipa-tion of, needs by the crop sector. Specific considerations for the other sectors have subsequently been add-ons. The crop sector also leads in terms of applications of, and capacities for, biotech. At any one time, the total content of active projects and associated R&D budgets are dominated by crops. The same pattern is seen in human resources. Generally, livestock comes second in these metrics, and forestry and aquaculture are often far behind—in many cases

---

[1] See: Current Status of the Water Efficient Maize in Africa (WEMA) Project « Biosafety Information Centre (biosafety-info.net)

disproportionate to the national contributions of these sectors. Moreover, the biotech content of investments in these two sectors is consistently bordering on the trivial. Training is a good example: the majority of training institutions specializing in forestry and aquaculture tend to focus on diploma and under-graduate level programmes intended to produce industry technicians but with limited attention to post-graduate training that can produce next-generation researchers and leaders. There are therefore a disproportionately low numbers of researchers and research facilities in these fields. Initiatives aimed at strengthening agricultural biotech applications in SSA countries must seek to understand this dynamic at the national level and develop strategies which bridge these gaps—leveraging on what is already in place or planned for the crop and livestock sectors. For example, there is need to exploit more deliberately the fact that most of the technologies, approaches and tools are generic. Countries should develop sharing arrangements across sectors.

References

Chambers JA, Zambrano P, Falck-Zepeda J, Gruère G, Sengupta D, Hokanson K (2014) GM agricultural technologies for Africa – A state of affairs. Report of a study commisioned by the African Development Bank and undertaken by The International Food Policy Research Institute (IFPRI). IFPRI, Washington, DC

FARA (Forum for Agricultural Research in Africa) (2011) Status of biotechnology and biosafety in sub-Saharan Africa: a FARA 2009 study report. FARA Secretariat, Accra

Gastrow M, Roberts B, Reddy V et al (2016) Public perceptions of biotechnology in South Africa. Report for the Public Understanding of Biotechnology Programme. South African Agency for Science and Technology Advancement, Pretoria. Available from: http://www.pub.ac.za/wp-content/uploads/2016/10/Public-Perceptions-to-Biotechnology.pdf. Accessed 20 Oct 2021

Juma C, Serageldin I (2007) Freedom to innovate: biotechnology in Africa's development: report of the high-level African panel on modern biotechnology. African Union and New Partnership for Africa's Development, Midrand

Kiome R (2015) A strategic framework for transgenic research and product development in Africa: report of a CGIAR study. ILRI, Nairobi

Lazarus JV, Ratzan SC, Palayew A et al (2021) A global survey of potential acceptance of a COVID-19 vaccine. Nat Med 27(2):225–228. https://doi.org/10.1038/s41591-020-1124-9

Makinde D, Mumba L, Ambali A (2009) Status of biotechnology in Africa: challenges and opportunities. Asian Biotechnol Dev Rev 11(3):1–10

Morris EJ (2011) Modern biotechnology – potential contribution and challenges for sustainable food production in sub-Saharan Africa. Sustainability 3(6):809–822

Nachega JB, Sam-Agudu NA, Masekela R, van der Zalm MM, Nsanzimana S, Condo J, Ntoumi F, Rabie H, Kruger M, Wiysonge CS, Ditekemena JD, Chirimwami RB, Ntakwinja M, Mukwege DM, Noormahomed E, Paleker M, Mahomed H, Muyembe Tamfum J-J, Zumla A, Suleman F (2021) Addressing challenges to rolling out COVID-19 vaccines in African countries. Lancet Glob Health. https://doi.org/10.1016/S2214-109X(21)00097-8

Nang'ayo F (2006) The status of regulations for genetically modified crops in sub-Saharan Africa. African Agricultural Technology Foundation, Nairobi

Ochieng JW, Ananga A (2019) Biotechnology in agricultural policies of sub-Saharan Africa. In: Elements of bioeconomy. IntechOpen. https://doi.org/10.5772/intechopen.8556

Printed in the United States
by Baker & Taylor Publisher Services